Series Editor's Foreword

Huge progress towards the understanding of biological systems and processes continues to be made through the application of the principles and techniques of organic chemistry. As a result, chemical biology now forms part of organic chemistry and biochemistry courses at University. This Oxford Chemistry Primer provides a concise introduction to chemical biology for chemistry and biochemistry students at the start of their University apprenticeships, and will serve to stimulate and excite their interest in this important area of science where chemistry overlaps with biology. As with other 'Foundation' Chemistry Primers this primer will also be of interest to young people studying chemistry or biology in their final year at school or college and their teachers.

Professor Stephen G. Davies
The Dyson Perrins Laboratory
University of Oxford

Preface

Chemical biology is a subject born out of a desire to understand the molecular basis of life. What are the molecules found in cells? How do their intrinsic properties equip them to perform all the complex processes found in living systems? This book introduces the fundamental chemistry of the molecules that are essential to all cells. The molecules we discuss include amino acids and sugar phosphate derivatives, and the macromolecules derived from them (proteins and nucleic acids, respectively), and the phospholipids and their derivatives that form the basis of biological membranes. In such a short text, it is not possible to provide a comprehensive account of the chemistry of these molecules; instead, this book attempts to introduce important concepts concerning their intrinsic chemistry. The aim is to provide the fundamental ideas relating to the chemistry of life that can then be applied in due course to more advanced aspects of chemical biology.

This book developed from a course of lectures and classes that the three of us taught together for several years to undergraduate chemists and

Chemical biology is a subject in which the principles of chemistry are applied to understand the function of biological molecules in their cellular environment.

The solution of research problems in chemical biology is the basis by which the subject develops. This type of research is leading both to an increased understanding of the molecular basis of life, and to exciting new applications of chemistry in subjects such as medicine and materials science.

This book uses specific case studies to illustrate key features of chemical biology. It is impossible to acknowledge all the scientists who have contributed to these studies, but they are the other inspiration of this book. Special mention should, however, be made of the pioneering research on globins by Max Perutz and colleagues which forms the basis of much of Chapter 4; and of the research on TIM by Jeremy Knowles and colleagues which is highlighted in Chapter 5.
The depictions of three-dimensional protein structures were produced using the program 'Molscript' (P. J. Kraulis (1991) *Journal of Applied Crystallography*, **24**, pp. 946–50).

biochemists at the University of Oxford. It was our students on this course, and a group of graduates and post-doctoral research assistants who acted as their mentors, inspired this book. They, and many others who have been exposed to parts of this material, have provided invaluable feedback for which we are very grateful.

We should also like to take the opportunity to thank specifically our colleagues who have provided detailed assistance in the production of the final version of book: Lorna Smith and Phillip Rendle were of great help in producing some of the diagrams involving macromolecular structures; Jack Heinemann kindly helped us avoid errors of genetics in Chapter 9; and Peter Steel provided wise advice on organic chemistry. The input of Ashley Sparrow and Claire Vallance, who criticized earlier drafts of the whole book from the perspective of non-specialists, was especially useful. The errors that remain are, of course, solely our responsibility. A. J. P. gratefully acknowledges funding from the University of Canterbury in the form of an Erskine Fellowship and study leave. Without such generous support, this book would have had an even more prolonged gestation period.

Cambridge   C. M. D.
Christchurch   J. A. G.
July 2001   A. J. P.

Contents

1 The chemicals of biological systems

1.1 Introduction

This book is concerned with the chemistry taking place in the cells of living organisms. Cells consist of a semipermeable membrane enclosing an aqueous solution rich in a diverse range of chemicals (Fig. 1.1). To the chemist, cells are, in essence, sophisticated machines that undertake a wide range of chemistry in an organized fashion. Cells have the potential to grow, replicate and produce closely related daughter cells, thereby handing down their controlled chemistry to the next generation. These remarkable characteristics all emerge from the chemical properties of the constituent molecules of cells.

The chemicals present in cells appear to have been selected by the processes of evolution for their chemical utility. The aim of this text is to show that many cellular processes can be understood in simple molecular terms. Although many biochemical molecules have complex structures, their biological properties can often be rationalized in terms of rather simple chemistry. A comprehensive account of the chemistry of biological systems is not the objective of this book; instead, a series of examples will be used to exemplify many of the principles that are important for understanding the chemistry of cells.

Figure 1.1 illustrates some gross features of a typical prokaryotic cell—a cell lacking a nucleus. All bacteria are prokaryotes, whereas all complex multicellular organisms such as plants and animals, as well as many unicellular species, are eukaryotes—their cells have a nucleus which houses DNA. Eukaryotic cells are rather more complex in structure and function; for mechanical strength they may utilize an internal skeleton in addition to, or instead of, an external cell wall; and they contain a range of discrete internal compartments. Although the detailed workings of prokaryotic and eukaryotic cells are different, the types of chemicals present, and the factors affecting their location, are similar: lipids, and other molecules with non-polar surfaces, are found in membranes; and polar entities such as sugars and amino acids are retained in aqueous solution.

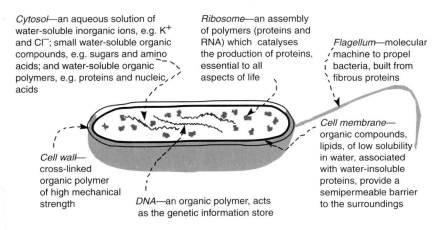

Cytosol—an aqueous solution of water-soluble inorganic ions, e.g. K^+ and Cl^-; small water-soluble organic compounds, e.g. sugars and amino acids; and water-soluble organic polymers, e.g. proteins and nucleic acids

Ribosome—an assembly of polymers (proteins and RNA) which catalyses the production of proteins, essential to all aspects of life

Flagellum—molecular machine to propel bacteria, built from fibrous proteins

Cell membrane—organic compounds, lipids, of low solubility in water, associated with water-insoluble proteins, provide a semipermeable barrier to the surroundings

Cell wall—cross-linked organic polymer of high mechanical strength

DNA—an organic polymer, acts as the genetic information store

Fig. 1.1 A chemist's schematic view of a bacterial cell.

1.2 The elemental composition of cells

After C, H, N and O, the other elements important, or essential, to life are B, Ca, Cl, Co, Cu, Fe, K, Mg, Mn, Mo, Na, Ni, P, S, Se, Si and Zn.

The composition of sea water has also been modified during the course of life on earth. The co-evolution of life and the earth form the basis of the 'Gaia' hypothesis that has been put forward and discussed by James Lovelock.

Elements which are abundant in the earth's crust, e.g. Al (8.2 per cent of the atoms) and Si (28 per cent of the atoms), but which are not readily soluble, are present at only low levels in sea water and often only present at low levels in cells (Fig. 1.2).

For elements lying close to the diagonal line in Fig. 1.2, the average concentration found in the human body is comparable with that found in sea water.

Iron is an example of an element more plentiful within cells than in the oceans. Iron carries out a diverse range of chemistry that is indispensable to cells. For over half of the 4.5 billion years of the earth's existence, its surface environment is believed to have been more highly reducing than at present. In particular, oxygen is thought to have accumulated to significant levels only about 2 billion years ago. Before the accumulation of oxygen, much of the iron on the surface of the earth was present as moderately soluble iron (II) salts. Once oxygen accumulated, however, more iron became trapped as iron (III) hydroxide that is very insoluble. As the availability of iron decreased, organisms evolved the ability to concentrate this element from their environment.

The contents of cells are related to, although different from, the chemical composition of their external environment. It is possible to rationalize the similarities and differences between cells and their environment in molecular terms. The use of particular chemicals by biology is related to their availability (now and in the past) and their chemical utility. The chemistry of cells is dominated by compounds made up of a small number of elements. For example, 99 per cent of the *atoms* are of four elements: hydrogen (62.8 per cent); oxygen (25.4 per cent); carbon (9.4 per cent) and nitrogen (1.4 per cent). This fact reflects the predominant role of water in cells. Indeed, as can be seen from Fig. 1.2, the composition of cells is related to the composition of sea water: in general, elements abundant in sea water are abundant in cells and vice versa. This is presumably because chemicals present in sea water were available during evolution. For example, as in sea water, many inorganic ions such as sodium, potassium and chloride are present at high levels in cells. The composition of sea water, in turn, reflects the chemicals available at the surface of the earth, modified by their water solubility.

The way in which the composition of cells differs from that of sea water sheds light on the chemistry of life. All cells must concentrate and retain foodstuffs and other essential chemicals. Cells must also discard unwanted chemicals into the environment. Some chemicals plentiful within cells are absent, or present at lower levels, in the sea. These chemicals are enriched in cells because of their chemical utility. In Fig. 1.2 the elements which are enriched in cells appear above the diagonal line, e.g. nitrogen, phosphorus and iron.

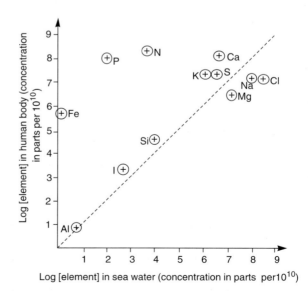

Fig. 1.2 Comparison of the concentrations of elements in the human body and in sea water.

Loan Receipt
Liverpool John Moores University
Library and Student Support

Borrower Name: Maduma,Ayanda
Borrower ID: **********1112

Foundations of chemical biology /
31111010509865
Due Date: 14/11/2011 23:59

Total Items: 1
24/10/2011 14:50

Please keep your receipt in case of
dispute.

1.3 The molecules present in cells

Many different organic compounds are found in cells. The interconversion of organic compounds is critical to the functioning of a cell—it provides both the chemical energy required to fuel the cell's activities and the materials needed by the cell to construct other molecular species. The reactions of chemicals within cells are collectively known as *metabolism* and so these organic compounds are known as *metabolites*.

Some of the small organic metabolites are used as building blocks of polymers synthesized and used by cells. These polymers include proteins, constructed from amino acids, and nucleic acids, which are derived from sugars and phosphate ions in combination with another class of organic compound, the heterocyclic bases. The properties of these polymers are one of the most distinctive chemical features of living cells.

Life is inextricably linked with water. The interior of a cell is an aqueous solution, rich in a variety of chemicals including simple inorganic species such as salts, small organic molecules, and a range of polymers derived from these molecules. This solution of water-soluble chemicals is enclosed by membranes comprised of molecules not freely soluble in water. The interaction of cellular molecules with water is crucial in determining their biological properties and provides a focus for much of this book.

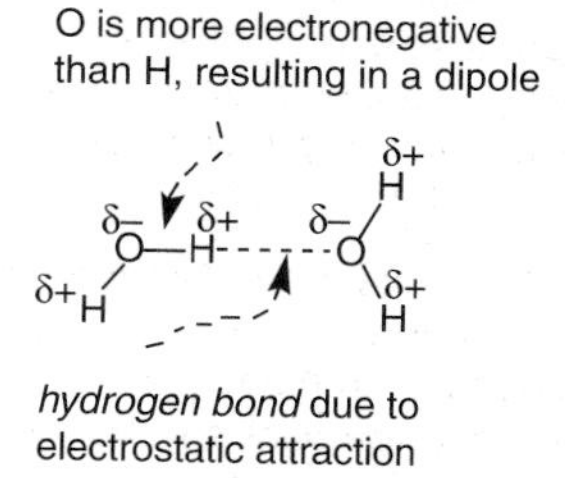

1.4 The importance of water

Because of their non-polar nature, most organic compounds cannot form hydrogen bonds with water molecules and so do not dissolve in aqueous solutions. Alkanes, for example, are immiscible with water and float on top of it. This arrangement minimizes the surface area of the organic compound in contact with water, leaving the water molecules free to hydrogen bond with each other. Molecules that have a very highly non-polar surface are relatively rare in biochemistry. Fats, e.g. triglycerides, are of this type (Fig. 1.3); they are immiscible with water and segregate themselves from the aqueous environment. Molecules, and portions of molecules, which prefer to avoid contact with water are termed *hydrophobic*.

The only organic compounds freely soluble in water bear polar groups on the carbon framework which *can* hydrogen bond with water. Hydroxyl groups fall into this category. Sugars, such as glucose, dissolve in water by virtue of the hydroxyl groups attached to the carbon framework (Fig. 1.4). Functional groups that interact favourably with water are termed *hydrophilic*.

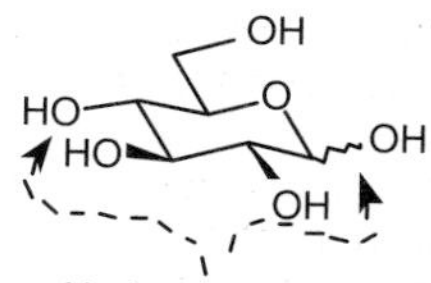

Table 1.1 Approximate chemical composition of a typical cell.

	Per cent of total cell weight
Water	70
Inorganic ions	1
Sugars	3
Amino acids	0.5
Nucleotides	0.5
Lipids	2
Macro- molecules	22

Hydrogen bonding results from an electrostatic attraction between an electron-deficient hydrogen atom and an electron-rich centre. When hydrogen is attached to an electronegative element, it becomes relatively positive and can interact favourably with relatively negative centres. For example, in water:

O is more electronegative than H, resulting in a dipole

hydrogen bond due to electrostatic attraction

'Hydrophobic' is derived from the Greek words 'hydro' for water and 'phobic' for fearing.

Hydroxyl groups can hydrogen bond effectively with water

Fig. 1.4 Glucose: an example of a water-soluble biochemical molecule.

'Hydrophilic' is derived from the Greek words 'hydro' for water and 'philic' for loving.

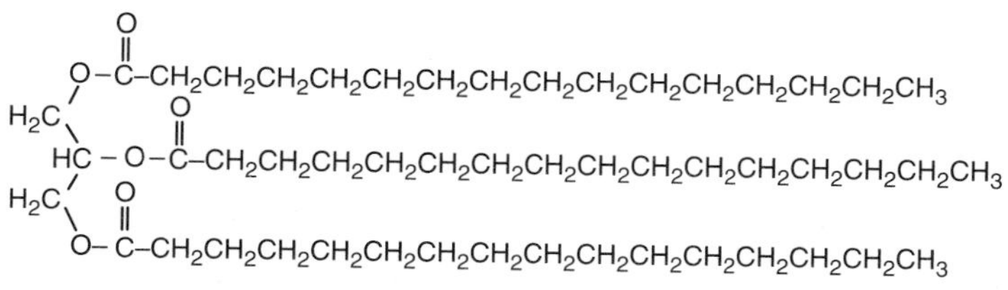

Fig. 1.3 A triglyceride: a water-insoluble biochemical molecule.

The interaction of compounds with water is an equilibrium phenomenon. It can be related to Gibbs free energy changes (ΔG); these, in turn, have enthalpy (ΔH) and entropy (ΔS) components:

$$\Delta G = \Delta H - T\Delta S$$

where T is the temperature. ΔH is a measure of changes in heat associated with a process, whereas ΔS is a measure of changes of the degree of disorder of a system. Favourable processes involve an overall decrease in free energy ($\Delta G < 0$) because of either the liberation of heat ($\Delta H < 0$) or an increase in disorder, ($\Delta S > 0$) or both.

Dispersing a non-polar organic compound in water would force the water to adopt a more ordered structure in an effort to retain as much hydrogen bonding as possible. Segregation of the organic compounds from water minimizes this unfavourable entropy effect.

Phosphates are another important class of ionic functional group found in many metabolites. These deprotonated forms of phosphoric acids also hydrogen bond readily with water. They are discussed extensively in the latter half of the book.

Figure 1.6 illustrates a single tetramer, ABCD, formed from four distinct monomer units. There are a total of 24 possible tetramers derived from combining a set of four different monomers ($4 \times 3 \times 2 \times 1 = 24$). Up to 256 tetramers (i.e. $4 \times 4 \times 4 \times 4 = 256$) are possible if any combination of such a family of monomers may be employed, corresponding to choosing any of the four monomers at each position, e.g. ABAD.

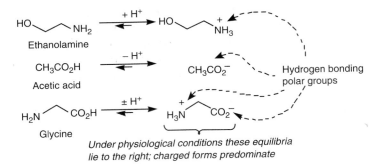

Under physiological conditions these equilibria lie to the right; charged forms predominate

Fig. 1.5 Representative water-soluble organic ions.

Many inorganic salts are soluble in water. Likewise, the introduction of charge into organic molecules enhances hydrophilicity. Charge arises in organic molecules primarily via acid–base chemistry, e.g. the protonation of amines to form ammonium salts. As examples (Fig. 1.5), ethanolamine, a biochemically important amine, is protonated at neutral pH, while acetic (ethanoic) acid is deprotonated. Both are charged at normal physiological pH. By analogy, amino acids, such as glycine, contain two opposite charges under the conditions found in cells.

The interactions of molecules with water are crucial in determining their biological functions. These interactions, in turn, are determined by the type, number and distribution of polar functional groups over the non-polar hydrocarbon backbone of a molecule.

1.5 Ordered molecular structures in biology

A key feature of many important biochemical molecules is that they adopt ordered structures. This ordering is the basis of their biological function. Two classes of ordering are highlighted in this text: the organized linking of monomers to form polymers, and the formation of ordered three-dimensional structures by some classes of biological molecule when they come into contact with water.

In biological polymers derived from a family of monomer units (notably proteins and nucleic acids) the monomers are covalently linked in a specific order in the final polymer chain, as illustrated schematically in Fig. 1.6.

Organic molecules, whether small or large, which contain both hydrophobic and hydrophilic portions, have the potential to adopt ordered structures in water. There is a driving force for such molecules to

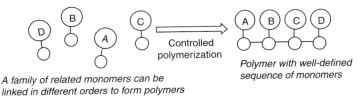

A family of related monomers can be linked in different orders to form polymers

Polymer with well-defined sequence of monomers

Fig. 1.6 A schematic representation of the ordering of monomers in biological polymers.

maximize the interaction of hydrophilic portions with water, whilst minimizing the exposure of hydrophobic regions. Proteins, nucleic acids and lipids all owe their biological function to the emergence of well-defined structures on interaction with water.

There are two different ways by which ordered structures emerge when biological molecules are in contact with water. A polymer chain can fold into a three-dimensional structure in which hydrophobic regions are buried away from the solvent, water (Fig. 1.7). This type of structure is found for most proteins and some nucleic acids (e.g. see Sections 3.6, 4.4 and 9.6).

Alternatively, molecules can associate non-covalently to form organized assemblies. This ordering is observed for both large and small molecules. Some biological polymers come together to form multimeric structures (Fig. 1.8). This association is important for the biological functioning of many proteins (e.g. see Sections 3.4 and 4.4). The DNA double helix (Section 9.9) also involves this type of molecular interaction. Lipids are small molecules rather than polymers. Their spontaneous association to form bilayer assemblies (Fig. 1.9) is the basis of biological membrane formation (Chapter 8).

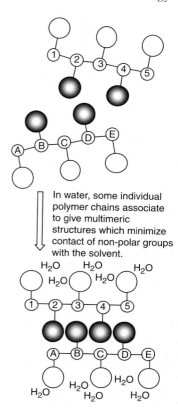

In water, some individual polymer chains associate to give multimeric structures which minimize contact of non-polar groups with the solvent.

Fig. 1.8 Schematic view of polymer chains associating to form a multimeric structure.

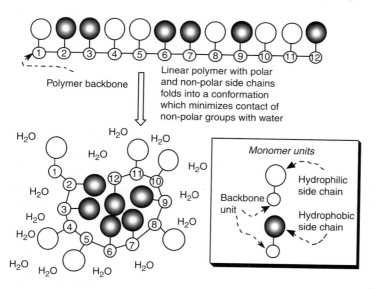

Polymer backbone

Linear polymer with polar and non-polar side chains folds into a conformation which minimizes contact of non-polar groups with water

Monomer units

Backbone unit

Hydrophilic side chain

Hydrophobic side chain

Fig. 1.7 Schematic view of a polymer folding into a well-defined shape in water.

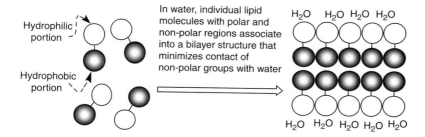

Hydrophilic portion

Hydrophobic portion

In water, individual lipid molecules with polar and non-polar regions associate into a bilayer structure that minimizes contact of non-polar groups with water

Fig. 1.9 Schematic view of lipids associating to form a bilayer.

Individual lipid molecules are *not* covalently linked together in membranes; the bilayer structure involves non-covalent assemblies of molecules. The same chemical principles are responsible for the adoption of these non-covalent assemblies as in the adoption of well-defined structures by proteins and nucleic acids. Individual lipid molecules contain both hydrophilic and hydrophobic regions. The bilayer structure allows the hydrophobic regions of lipid molecules to be buried away from water, leaving only the hydrophilic portions exposed.

A great deal of contemporary research in chemistry is associated with trying to understand and learn from biology. The study of simple chemical systems which mimic the chemistry found in complex biological systems is an important area of research known as *biomimetic chemistry* (e.g. see Section 4.5). It is useful in unravelling some of the basic principles underlying the chemistry found in cells.

1.6 An overview of the book: the emergence of biological function

Much of this book discusses the way in which small molecules present in cells combine to generate ordered structures with biological function (Fig. 1.10). To this end, the small molecules, and their derivatives, are considered in two groups. The first half of the book (Chapters 2–5) introduces one family of molecules: amino acids and proteins. The second half (Chapters 6–9) deals with molecules derived from sugars and phosphates, including the nucleic acids. In each half of the book, the individual components are described and then the discussion is elaborated to illustrate how biological function is facilitated as these individual units come together to form complex entities.

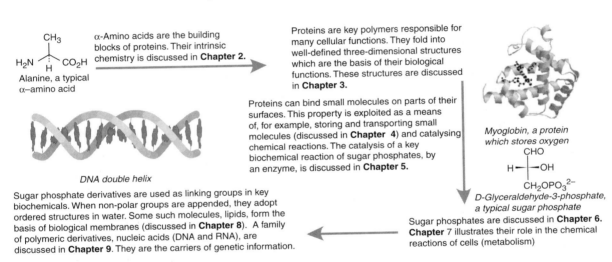

α-Amino acids are the building blocks of proteins. Their intrinsic chemistry is discussed in **Chapter 2.**

Alanine, a typical α–amino acid

Proteins are key polymers responsible for many cellular functions. They fold into well-defined three-dimensional structures which are the basis of their biological functions. These structures are discussed in **Chapter 3.**

Proteins can bind small molecules on parts of their surfaces. This property is exploited as a means of, for example, storing and transporting small molecules (discussed in **Chapter 4**) and catalysing chemical reactions. The catalysis of a key biochemical reaction of sugar phosphates, by an enzyme, is discussed in **Chapter 5.**

Myoglobin, a protein which stores oxygen

D-Glyceraldehyde-3-phosphate, a typical sugar phosphate

DNA double helix

Sugar phosphate derivatives are used as linking groups in key biochemicals. When non-polar groups are appended, they adopt ordered structures in water. Some such molecules, lipids, form the basis of biological membranes (discussed in **Chapter 8**). A family of polymeric derivatives, nucleic acids (DNA and RNA), are discussed in **Chapter 9**. They are the carriers of genetic information.

Sugar phosphates are discussed in **Chapter 6**. **Chapter 7** illustrates their role in the chemical reactions of cells (metabolism)

Fig. 1.10 A schematic overview of the book.

The science of new materials is one area which benefits from the principles learned from biology. As an example, one goal of current chemistry is to make molecular devices, where individual molecules undertake a role normally performed by a large-scale machine. 'Molecular machines' have been found to play an important role in biology (e.g. see Chapter 8). Such research into 'nanotechnology' is hoped to provide new levels of miniaturization for future technological applications.

Further reading

An overview of cell biology is given in: H. Lodish, D. Baltimore, A. Berk, S. L. Zipursky, P. Matsudaira and J. Darnell (1995) *Molecular Biology of the Cell*, 3rd edn, W H Freeman & Co Ltd, Oxford.

Bioinorganic chemistry is discussed in: P. A. Cox (1995) *The Elements on Earth*, Oxford University Press, Oxford; P. C. Wilkins and R. G. Wilkins (1997) *Inorganic Chemistry in Biology*, Oxford University Press, Oxford; and R. J. P. Williams and J. J. R. Frausto da Silva (1997) *The Natural Selection of the Chemical Elements*, Oxford University Press, Oxford.

Good recent general biochemistry texts include: C. K. Mathews, K. E. van Holde and K. G. Ahern (2000) *Biochemistry*, 3rd edn, Benjamin/Cummings, San Francisco; and R. H. Garrett and C. M. Grisham (1998) *Biochemistry*, 2nd edn, Sanders College Publishers, Fort Worth.

2 Introduction to amino acids and proteins

2.1 Overview: what are proteins and why are they special?

Proteins are biological polymers play important roles in virtually all the chemical processes of life. As such, they are abundant in all cells, representing approximately 15 per cent of the total cell mass. There are thousands of different types of proteins in even simple cells, but all proteins are derived from the same basic building blocks: a set of amino acids. An understanding of the chemistry of these amino acids, and how their properties change when they are polymerized in a protein, is therefore of key importance in the analysis of the *molecular* basis of the biological properties of proteins.

Proteins are linear polymers derived from α-amino acid monomers (see Figs 2.1 and 2.2). These α-amino acids are carboxylic acids with an amino group, a hydrogen atom, and a further substituent (R) attached to the α-carbon (i.e. the carbon adjacent to the carboxyl group). In proteins, the substituent R is limited to one of 20 possible groups (or occasionally a derivative of one of these groups) and the individual amino acids are joined via amide linkages to form a polypeptide chain. An amide linkage is the result of the condensation of the amine and carboxylic acid functional groups of adjacent monomers; the remaining portion of one monomer is termed an amino acid *residue*. A protein is typically a polypeptide chain of several hundred such residues. Shorter polymers are frequently called *peptides*.

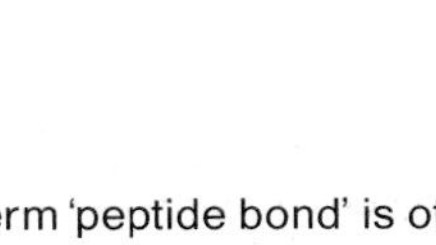

Fig. 2.1 Generalized structure of an α-amino acid. In aqueous solution these occur as 'zwitter-ions' (see Section 2.2).

The term 'peptide bond' is often used to signify the amide bond between α-amino acids in peptides and proteins.

In this book, bonds designated $\sim\sim$ indicate a continuation of the polymer chain.

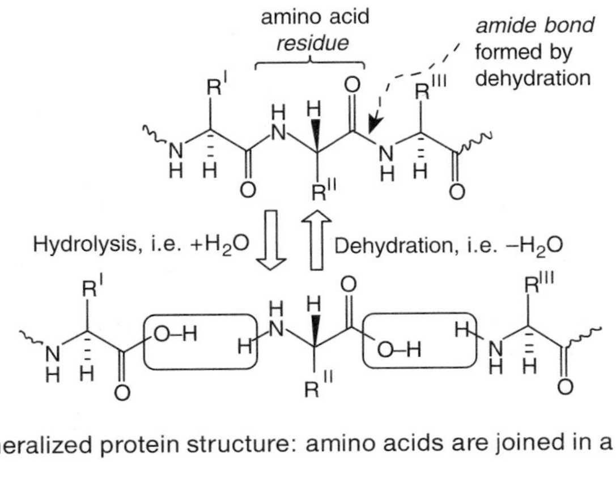

The term 'peptide' is generally used to signify a polymer with less than about 30 residues. For simplicity, in this book the term protein is used for all biological polypeptides regardless of their size.

Fig. 2.2 Generalized protein structure: amino acids are joined in a linear chain.

Nylon-6 is a polymer of
6-aminohexanoic acid.

Proteins are not the only linear polymers of amino acids, e.g. nylon-6 is a synthetic analogue. A distinguishing feature of proteins is that they are composed of a variety of monomer units linked together in a defined sequence, i.e. not just a repeated pattern of one or two simple monomers as found in a synthetic polymer such as nylon. The huge variety of possible proteins underpins the diverse roles for which they are suited: as examples, some proteins act in a structural capacity, e.g. collagen and keratin (found in tendons and hair, respectively) function in biology as fibres; others, such as haemoglobin, are involved in transport of molecules within organisms; and yet more, *enzymes*, act as catalysts for the essential chemical reactions of cells.

It is the presence of a range of different functional groups (R^I, R^{II}, R^{III}, etc. in Fig. 2.2) attached to the α-carbon of each amino acid residue that provides the structural and chemical versatility essential to protein function. These functional groups branch off the polymer backbone, the *main chain*, and are termed *side chains*. Proteins are generated in biological systems by controlled, sequential addition of monomer units. The specific amino acids present in a protein, and their order within the chain, give rise to the distinctive shape and chemical functionality of each final polymer.

In order to understand the structures and functions of proteins, the chemistry of the functional groups present in the protein will be examined. Proteins usually adopt highly ordered structures in the cellular environment, based on the nature of the substituents on the polymer chain (polar and non-polar) and on the stereochemical features of both the amino acids and the peptide bonds that link them. An appreciation of these structural principles will be used in analysing case studies of well-studied proteins in later chapters.

The importance of polar and non-polar substituents in determining the structures that polymers adopt in aqueous solution was outlined in Section 1.5.

2.2 The acid–base chemistry of α-amino acids and proteins

An understanding of the structures and simple chemical features of α-amino acids is important for appreciating the properties of proteins. Acidity and basicity of amino acids, and hence of proteins, is the chemistry which will be discussed in most detail. These are important chemical properties in biological systems. Simple acid–base chemistry will be illustrated using glycine, the simplest amino acid. Initially, the properties of free glycine will be described, followed by a discussion of the way in which these properties are modified when glycine is incorporated into proteins.

Acidity reflects the tendency for a bond to break and generate a proton. Basicity reflects the reverse process: the tendency to form a bond between a proton and an atom with an available pair of electrons. The ease with which these processes occur for particular chemical functional groups is quantified in terms of the equilibrium constant, K_a, for the reactions. A general reaction of this type in which a functional group HA, the *conjugate acid*, dissociates to give a proton and A⁻, the *conjugate base*, is

$$H - A \rightleftharpoons H^+ + A^-$$

K_a is the ratio of concentrations (a more precise analysis uses 'activities' rather than concentrations) of products to starting materials, i.e.

$$K_a = [H^+][A^-]/[HA]$$

2.2.1 Glycine

The simplest amino acid is glycine, in which two hydrogens are attached to the α-carbon, i.e. the substituent, R in Fig. 2.1, is hydrogen. The overall chemical properties of such a molecule reflect the combination of the properties of the various *functional groups* of the molecule, each of which has a distinctive chemistry.

Glycine exists primarily as charged species in aqueous solution as explained later (see Fig. 2.5).

Glycine

In glycine, the only functional groups attached to the hydrocarbon framework are an amino group and a carboxyl group. Each of these groups can exist in different protonated forms, depending on the pH.

In strong acid solution the carboxyl group will be present in the free acid form (i.e. as a CO_2H group). This group is acidic as it can lose a proton, in response to the addition of base, to form a carboxylate ion.

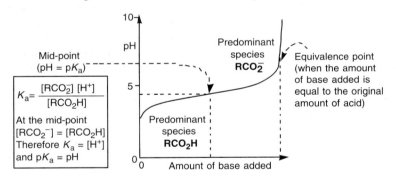

The change of pH as base is added to an aqueous solution of a typical carboxylic acid is shown in Fig. 2.3. At pH values above ca 5 (the pK_a of the acid), the carboxylate form predominates. In this case, the conjugate base is a charged species, and the acidity of the carboxylic acid group reflects the stability of this anionic species relative to the uncharged molecule in aqueous solution.

The characteristic features of a base are the presence of a non-bonded electron pair and a relatively high stability of the corresponding protonated form. The other functional group of glycine is an amine which contains a nitrogen atom with a non-bonded pair of electrons.

Mid-point
(pH = pK_a)

$$K_a = \frac{[RCO_2^-][H^+]}{[RCO_2H]}$$

At the mid-point
$[RCO_2^-] = [RCO_2H]$
Therefore $K_a = [H^+]$
and $pK_a = pH$

Predominant species RCO_2^-

Predominant species RCO_2H

Equivalence point (when the amount of base added is equal to the original amount of acid)

pH

Amount of base added

Fig. 2.3 Titration curve for a typical carboxylic acid.

RNH₂ predominates

Mid-point
(pH = pK_a)

Equivalence point

$\overset{+}{R}NH_3 \rightleftharpoons \overset{+}{H} + RNH_2$

pH

RNH_3^+ predominates

Amount of base added

Fig. 2.4 Titration curve for a typical amine.

The acidity of carboxylic acids is due to the stability of the carboxylate anion relative to the acid. The charge of the carboxylate resides, primarily, on two equivalent electronegative oxygens; each oxygen has to accommodate only half a negative charge.

The equivalence of the two oxygens in the carboxylate anion is not clear in a single structure comprised of simple bonds and charges. This problem can be addressed by imagining the overall structure as a weighted average of simple structures (*canonical* or *resonance* forms). Here two structures are drawn, corresponding to the charge residing on either of the two oxygens. These structures are equivalent and have identical energies. The 'real' structure is envisaged as the 'average' of these extreme representations (*not* a rapidly interconverting mixture of two discrete structures).

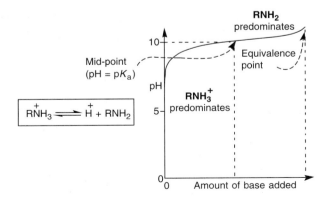

The relationship between these two canonical forms can be shown using 'curly arrows' which are used in organic chemistry to denote the movement of pairs of electrons.

The negative charge is shared.

Acidity is quantified in terms of pK_a values. At the mid-point of a titration, an acid, HA, is half deprotonated, the concentrations of HA and A⁻ are equal, and the equilibrium constant $K_a = [H^+]$. The equilibrium constants for different acid–base reactions have widely differing values. By analogy to pH, logarithms are used to compress the scale. Since most acids of interest are only partly dissociated, the sign of the logarithm is changed so that most values are positive numbers:

i.e. $\log K_a = pK_a$

The *smaller* the value of the pK_a, the *stronger* the acid.

Some approximate pK_a *values*:
Alcohols, ROH, 17
Carboxylic acids, RCO_2H, 4–5
Thiols, RSH, 8
Ammonium ions, RNH_3^+, 10.

A zwitterion is a neutral molecule that carries equal numbers of oppositely charged functional groups. Zwitterions have many of the properties of salts.

pK_a values vary somewhat, depending on the precise molecular structure and the environment in which the acid–base chemistry is taking place.
For example, the carboxylic acid group of glycine is more acidic than a simple carboxylic acid ($pK_a = 2.35$ rather than ca 4–5) since the corresponding carboxylate anion is more stable in the presence of the adjacent, positively charged, ammonium ion. The effect of the environment on pK_a is particularly important in non-polar conditions, such as in the interior of proteins.

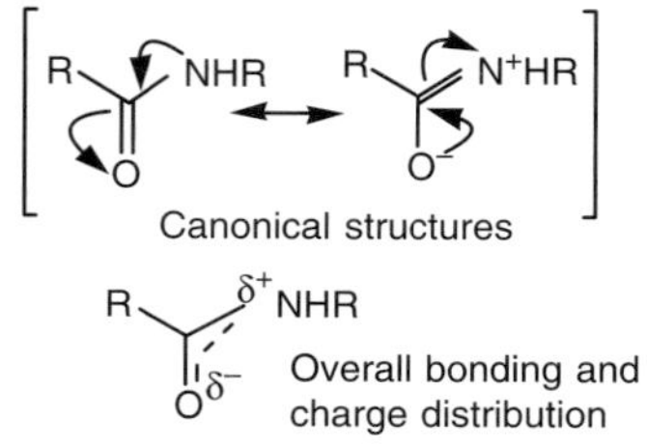

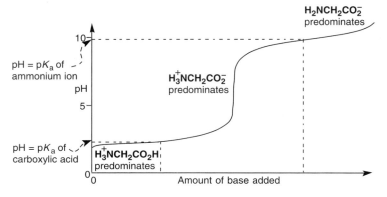

Fig. 2.5 Titration curve for glycine.

The conjugate acid of an amine, a substituted ammonium ion, is relatively stable in aqueous solution. Such an ion in aqueous solution can be deprotonated, but less readily than a carboxylic acid; the pK_a is correspondingly higher. The acid–base equilibrium and titration curve is shown in Fig. 2.4. Only at a pH above ca 10 (the pK_a of the substituted ammonium ion) does the amine exist predominantly in the uncharged form.

Since glycine contains both an amine and a carboxylic acid moiety, its titration behaviour incorporates both features, as can be seen in Fig. 2.5. The pK_a values of the carboxylic acid and ammonium groups of glycine are 2.35 and 9.78, respectively. In neutral aqueous solution (pH 7) both the functional groups of glycine are charged (Fig. 2.5). Therefore, glycine and other amino acids are really ammonium carboxylate 'zwitterions' rather than 'amino acids'; subsequent structures will reflect this.

2.2.2 The chemistry of amides: polymers of glycine

What are the acid–base properties of glycine when incorporated into a polypeptide? When the simple dimer, glycyl-glycine, is compared with its constituent monomers, the ammonium group of one glycine and the carboxylate group of the other are still present in the dimer, and their acid–base chemistry is little affected. The other ammonium and carboxylate groups are no longer present, since they are linked to form an amide bond.

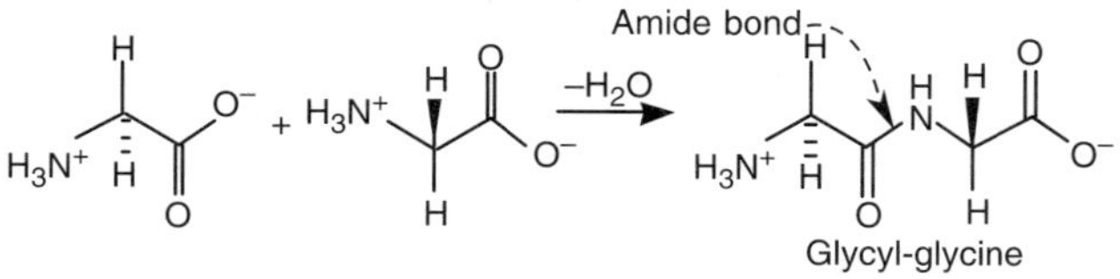

Fig. 2.6 The structure and bonding of amides.

By contrast with the nitrogen atom of the amine, the amide nitrogen is essentially non-basic. There is an additional interaction between the nitrogen and the unsaturated carbon atom, resulting in partial double-bond character. This can be depicted using canonical structures (Fig. 2.6), in an analogous fashion to those drawn to explain the special stability of the carboxylate anion.

When a nitrogen is positioned adjacent to an unsaturated carbon in a molecule, it is generally found to adopt a trigonal planar, rather than pyramidal, shape. By adopting this shape, the C–N bond is more stable: the electrons are delocalized in a partial double bond.

As well as being weak bases, amides are also only weak acids. Deprotonation of an amide produces an anionic conjugate base. This is similar in structure to a carboxylate ion, but with one of the oxygens replaced by a nitrogen atom. Since nitrogen is less electronegative than oxygen, this species is less stable than a carboxylate ion, and hence harder to form.

The fact that amides are neither strong acids nor bases is important for the chemistry of proteins. The overall acid–base properties of the main chain of a typical protein are essentially the same as those of glycine: there is an ammonium group at one end and a carboxylate group at the other; the intervening main chain, comprised of amide linkages, plays essentially no part in the acid–base chemistry of the final molecules. The chemical potential of proteins is not focused in the main chain; instead it is the range of chemical groups found in the side chains of the other amino acids which gives rise to the varied properties of proteins.

2.3 A survey of the α-amino acids of proteins

The proteins found in all living organisms are derived from a repertoire of 20 α-amino acids: glycine and 19 others. Not all proteins contain all 20 amino acids, although most do, and in certain cases modifications to some of these amino acids can occur. However, with a few exceptions, the 20 α-amino acids provide the building blocks for proteins from whatever source. This reflects the universal nature of the genetic code (see Section 9.11). It is important to have a thorough appreciation of the side chains of the amino acids found in proteins. These side chains are not special to biology, but include many of the simple functional groups familiar from organic chemistry.

2.3.1 Amino acid residues with hydrocarbon side chains

The side chains of several of the amino acids in proteins are simple hydrocarbons. Thus, when they are incorporated into a protein chain, no additional functional groups are introduced into the molecule. The non-polar nature of these residues, which precludes hydrogen bonding, plays a critical role in the structure of proteins, as will become apparent later: these residues are hydrophobic.

alanine residue   valine residue   isoleucine residue   leucine residue   phenylalanine residue

When the substituent is a methyl group, the amino acid is alanine. The remaining side chains in this class are either branched hydrocarbon chains (*i*-propyl (in valine) and two butyl isomers (in leucine and isoleucine)) or 'aromatic' groups as in phenylalanine.

The low basicity of amides can be attributed to the fact that the extra electrons on nitrogen are not a non-bonded lone pair, but involved in a favourable bonding interaction (see Fig. 2.6); protonation of the nitrogen atom destroys this interaction. Under strongly acidic conditions (e.g. conc. H_2SO_4) amides actually protonate on oxygen rather than nitrogen, since the *oxygen* of an amide *does* possess non-bonded electrons. Protonation of the amide oxygen catalyses the reaction of nucleophiles with amides. Hence, when polypeptides are heated in the presence of strong aqueous acid, they hydrolyse. (The hydrolysis of amides is discussed further in Section 7.2.)

The breakdown of proteins during digestion is helped by the low pH environment found in the stomach. 'Indigestion' can result from an excess of acid, and most treatments for this condition aim to neutralize this excess.

The methyl group of alanine is the only unbranched hydrocarbon side chain found among the protein amino acids. Simple homologues, such as ethyl or *n*-propyl, are not found.

Because of the carboxylic acid function, aspartic and glutamic acids are often termed 'acidic' residues. This terminology can be confusing: since they are usually found in proteins in the conjugate base form, the side chains of these amino acids often act as bases in biochemistry (see Chapter 5).

For amino acid side chains which are involved in acid–base equilibria, the acid or base form which predominates at physiological pH is boxed in the structures which accompany this section of the text.

The titration curves for aspartic and glutamic acid residues resemble those of simple carboxylic acids (cf. Fig. 2.3). Note that the exact pK_a of functional groups varies a little depending on the environment (either in response to other functional groups in the molecule, or in the surrounding solvent, etc).

2.3.2 Amino acid residues with carboxylic acid side chains

The side chains of two of the amino acids, aspartic and glutamic acids, contain carboxylic acid functional groups linked by a hydrocarbon spacer, of one or two methylene groups respectively, to the α-carbon. The side chain of each of these groups behaves as a simple carboxylic acid with a pK_a of approximately 4–5. At neutral pH, these groups will, therefore, be present in the anionic conjugate base form (aspartate and glutamate).

aspartate residue   glutamate residue

2.3.3 Amino acid residues with amide side chains

A further two amino acids, asparagine and glutamine, are closely related to aspartic and glutamic acids. In these, instead of a carboxylic acid, the side chain contains an amide group. Amides can participate in hydrogen bonding, but they are neither strong acids nor bases, and do not affect the acid–base chemistry of proteins.

asparagine residue   glutamine residue

The amino acids in this section are often said to have 'basic' side chains. This reflects the fact that they are readily protonated. If they are present in a protein in an unprotonated form, they can act as bases. If they are present in a protonated form, they can act as acids.

2.3.4 Acyclic amino acid residues with basic nitrogen-containing side chains

Two of the protein amino acids have side chains consisting of a linear carbon chain terminating in a basic nitrogen functional group. The side chain of lysine is a four-carbon chain ending in an amino group. This primary amine bears a non-bonding electron pair and is of similar basicity to the amines considered previously. The pK_a of the corresponding ammonium ion is 10.5 and, at neutral pH, this group is present in solution as a cation.

lysine residue   arginine residue

It is a useful exercise to draw canonical structures of the protonated form of arginine to illustrate the dispersal of charge over all three side chain nitrogens.

In the case of arginine, protonation of the basic nitrogen leads to a cation in which the positive charge is dispersed over *three* nitrogen atoms. This factor ensures an enhanced stability to the protonated form of arginine which has a pK_a of 12.5 and is present as a cation under physiological conditions.

2.3.5 Amino acid residues with hydroxyl functional groups

The side chains of three amino acids contain hydroxyl groups. Serine and threonine are simple alcohols. For each of these residues, the hydroxyl group is attached to a carbon adjacent to the α-carbon. Threonine is distinguished from serine by an extra methyl group that makes it a secondary, rather than a primary, alcohol. An isolated hydroxyl group can act as an acid or a base, but neither process is especially favourable (the pK_a of the hydroxyl of serine is approximately 16).

In tyrosine, the hydroxyl function is attached to an aromatic ring. Here the functional group is a phenol. The aromatic ring stabilizes the charge on the deprotonated form. This enhances the stability of the conjugate base and lowers the pK_a (to ca 10) facilitating acid–base chemistry. Tyrosine is usually found in the hydroxyl form, but it is occasionally found to act as an acid under physiological conditions.

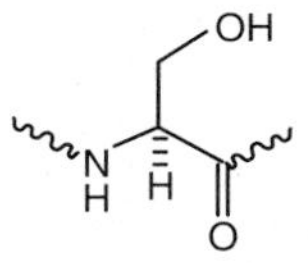

serine residue

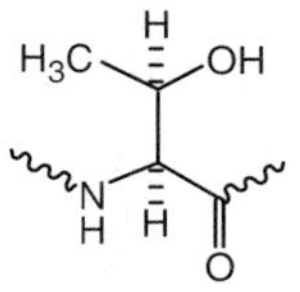

threonine residue

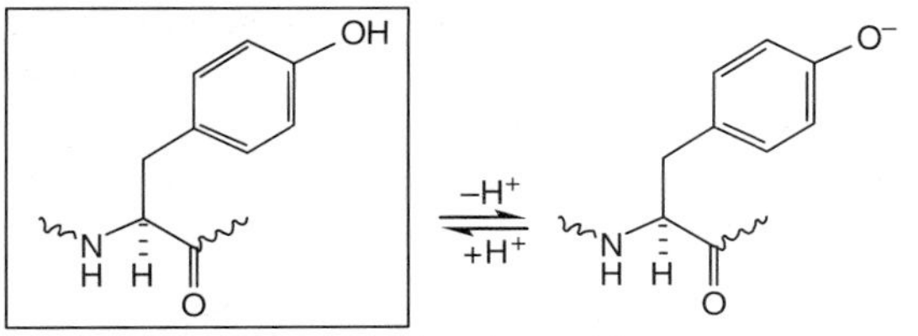

tyrosine residue

The negative charge of the deprotonated form of tyrosine can be dispersed over three of the carbons of the aromatic ring. Convince yourself about the charge distribution of the conjugate base of tyrosine by drawing canonical structures.

2.3.6 Sulphur-containing amino acid residues

The side chains of two protein amino acids have sulphur-based functional groups. Cysteine is the sulphur analogue of serine, containing a thiol functional group rather than a hydroxyl function. In aqueous solution such groups are moderately acidic (pK_a ca 8). However, the properties of sulphur differ from those of oxygen and thiols do not form strong hydrogen bonds. In general, sulphur-containing side chains behave as relatively non-polar groups. In addition, the thiol group has unique chemical properties: it is the most readily oxidized of all the functional groups under consideration (Fig. 2.7). When two thiols are oxidized, a disulphide bond results. Disulphide bonds are important features of some protein structures and are considered in Chapter 3.

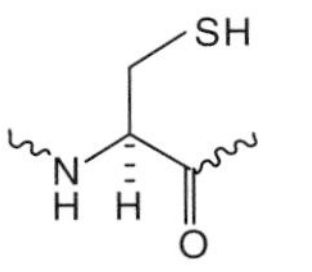

cysteine residue 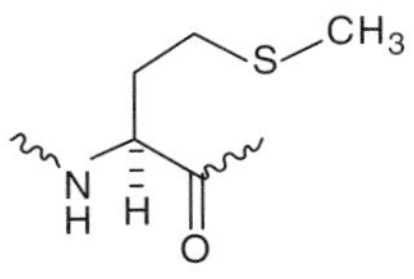   methionine residue

The amino acid methionine contains a thioether group rather than a thiol. For the present discussion, the most significant feature of the methionine side chain is its generally non-polar character.

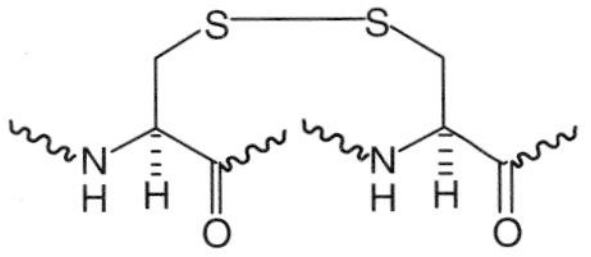

Reduction $+2H^+ + 2e^-$ Oxidation $-2H^+ - 2e^-$

Disulphide-linked residues, known as a cystine residue

Fig. 2.7 Redox chemistry of cysteine.

A heterocyclic compound contains a ring in which one of the ring atoms is not carbon.

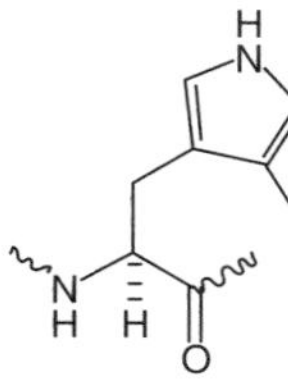

tryptophan residue

If both protonated and non-protonated forms are present, then the side chain can act as both an acid and a base. This is the situation for histidine. For this reason, histidine residues are often important in catalysing biochemical acid–base processes. The special chemistry of histidine is important in the functioning of many proteins. Two of the succeeding chapters describe proteins which have essential histidine residues: the globins in Chapter 4, and triose phosphate isomerase in Chapter 5.

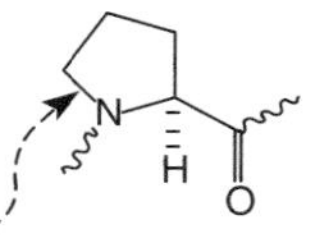

proline residue

The side chain links back and connects to the α-nitrogen, forming a ring.

Tetrahedral arrangement of four single bonds to carbon:

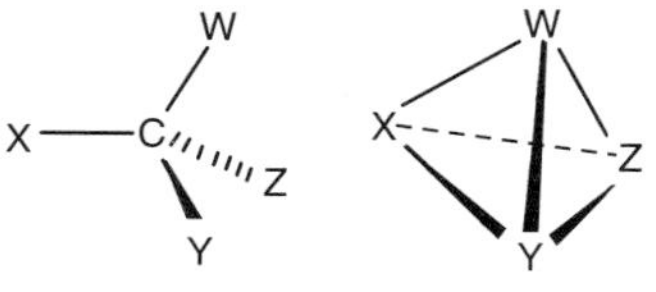

2.3.7 Amino acids containing nitrogen heterocycles

The final three protein amino acids are rather different from one another, but they each contain cyclic structures involving nitrogen which are responsible for their distinctive chemistry.

Tryptophan has a nitrogen embedded in a large aromatic framework (an indole) which behaves as a non-basic nitrogen, although it can form hydrogen bonds. Tryptophan is more like the side chain of phenylalanine than most of the remaining nitrogen-containing side chains. It is a hydrophobic residue.

Histidine also has a side chain with an aromatic ring. In this case the ring (an imidazole) has two nitrogen atoms, and can be protonated. The charge on the cation of the protonated form is dispersed over the two nitrogen atoms. Histidine is moderately basic with the pK_a of the conjugate acid being ca 7. Such a pK_a allows both conjugate acid and base forms to be readily accessible at neutral pH. Histidine is ideally placed to act as an acid–base catalyst in proteins operating at around pH 7.

Protonated and neutral forms of histidine are both physiologically important. This evenly balanced equilibrium underpins histidine's key role in biological acid–base chemistry.

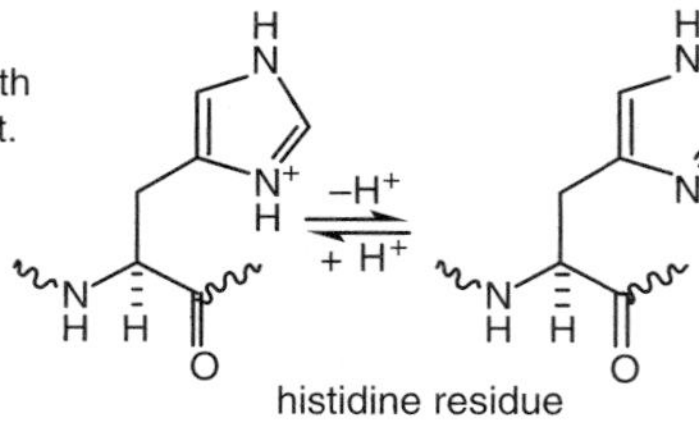

histidine residue

Finally, proline is fundamentally different from the other protein amino acids. The side chain comprises a three-carbon chain and, as with other hydrocarbon side chains, contributes no unusual chemical features to the amino acid or derivatives. In this case, however, a ring structure is formed by the side chain linking back to the α-amino nitrogen. This cyclic structure constrains the shapes which this amino acid can adopt. As a result, the presence of proline in a protein has significant effects on its three-dimensional structure (see Section 3.3).

2.4 The stereochemistry of α-amino acids

Stereochemistry is concerned with the shape of molecules. The properties of molecules are strongly influenced by their stereochemistry. An understanding of the stereochemical features of amino acids is essential for an appreciation of the fundamental principles of the structures of proteins.

All amino acids except glycine have four different substituents attached to the α-carbon. There are two distinct ways in which these substituents can be arranged in three-dimensional space. These two forms, *configurations*, are mirror images of each other. All amino acid residues in natural proteins have the same absolute configuration. These properties are illustrated in Fig. 2.8 for a typical amino acid, serine.

In serine the α-carbon is surrounded by four different groups: an ammonium, a carboxylate, a hydroxymethyl and a hydrogen atom. The

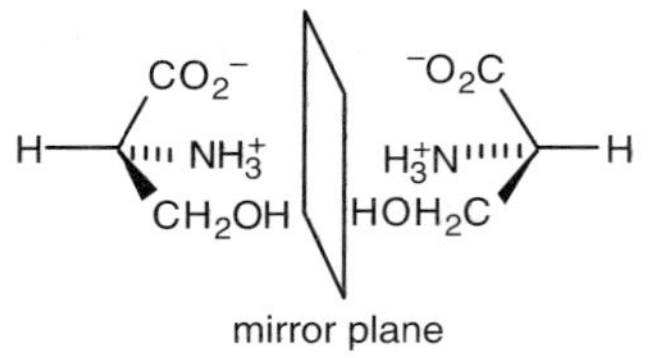

Fig. 2.8 The enantiomers of serine.

bonds to these substituents point towards the apices of a tetrahedron. Different configurations related in this mirror image fashion are known as *enantiomers* and the α-carbon of serine is a *chiral centre*.

Emil Fischer undertook pioneering research on the stereochemistry of natural products. In 1891, he reported the stereochemistry of a range of sugars. He introduced representations (so-called Fischer projections) to show the stereochemical relationships between related natural products. For these representations, Fischer compared the stereochemistry of the compound of interest with that of L- and D-glyceraldehyde.

Fischer carried out a great deal of research on the amino acids found in proteins. He synthesized serine in 1902 and resolved synthetic serine into the two enantiomeric forms in 1906.

2.4.1 The L and D stereochemical notation

It was known that all amino acids found in proteins have the same chirality, long before it was possible to tell which of the two absolute configurations was present in this class of molecules. In early stereochemical studies, the configuration of amino acids was related to known materials by a process of chemical correlation. One of the important reference compounds for stereochemical studies was glyceraldehyde. Glyceraldehyde has a single chiral centre; two enantiomers are possible, both of which were known at that time; Emil Fischer introduced the terms L and D to describe the two forms. The groups about the chiral centres of serine and glyceraldehyde are closely related—both have hydroxymethyl and hydrogen substituents. If the amino and carboxyl functional groups of serine are considered to be analogous to the hydroxyl and aldehyde groups, respectively, of glyceraldehyde, then serine (derived from a protein) has the same absolute configuration as L-glyceraldehyde. Since the structures of the other chiral amino acids from proteins are analogous to serine, all these amino acids are said to have the L-configuration.

L-glyceraldehyde D-glyceraldehyde L-serine D-serine

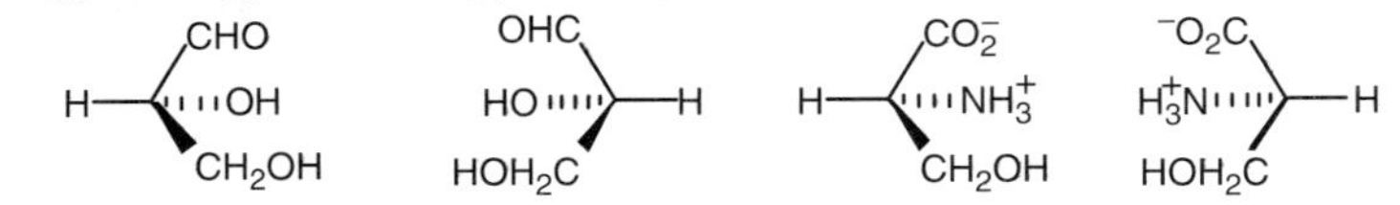

L-amino acid D-amino acid

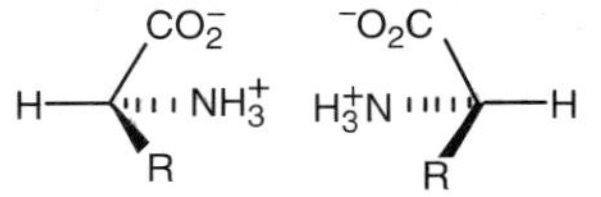

2.4.2 The Cahn–Ingold–Prelog rules for describing stereochemistry

The L and D terminology is a convenient notation for the stereochemistry of simple amino acids. However, it becomes difficult and subjective to use in other molecules when the analogies that have to be invoked are more tenuous than those so far described. In addition, some important molecules, such as the amino acid threonine, have more than one chiral centre. A wide variety of chiral compounds is now known, and the absolute configuration of chiral centres within these molecules can be determined using X-ray crystallographic methods. To deal with this situation, chemists have devised a notation, known as the *Cahn–Ingold–Prelog* (C-I-P) rules, to describe unambiguously the absolute configuration about any chiral centre.

Make molecular models of the two enantiomers of serine to convince yourself that the two structures are not identical.

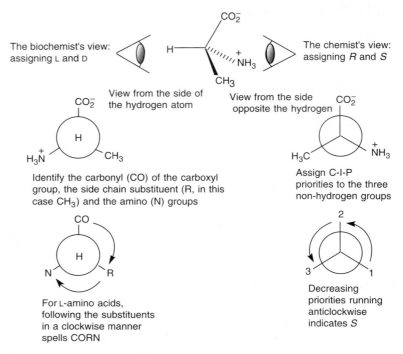

(i) *Priorities*
Decreasing priorities, numbered 1–4, are assigned to the atoms attached to the chiral centre. Higher priority is given to atoms of higher atomic number. In the case of all chiral amino acids found in natural proteins, this leads to the priorities $N > C$ (carboxyl), C (side chain) $>$ H. The amino group is numbered 1 and the hydrogen atom 4. When, as here, consideration of these atoms leaves an ambiguity, it is resolved by extending the rules to the next attached atoms. The priority of these atoms is assigned, as before.

(ii) *Translating priorities into descriptions*
The chiral centre is viewed from the side opposite the lowest-priority ligand. If the direction of decreasing priority (1, 2, 3) of the other three groups is clockwise, the configuration is designated *R*, whereas an anticlockwise ordering is designated *S*.

The biochemist's view: assigning L and D

The chemist's view: assigning *R* and *S*

View from the side of the hydrogen atom

Identify the carbonyl (CO) of the carboxyl group, the side chain substituent (R, in this case CH_3) and the amino (N) groups

For L-amino acids, following the substituents in a clockwise manner spells CORN

View from the side opposite the hydrogen

Assign C-I-P priorities to the three non-hydrogen groups

Decreasing priorities running anticlockwise indicates *S*

Fig. 2.9 Identifying the configuration of amino acids: alanine as an example.

The C-I-P rules involve assigning 'priorities' to the groups attached to the chiral centre and then relating these priorities to a description of the chiral centre. The two possibilities are known as *R* and *S* configurations. As an example, these rules are illustrated for alanine in Fig. 2.9.

In the case of alanine, the priorities of the atoms attached to the α-carbon are $N > C > H$. Hence the amino group is numbered 1 and the hydrogen 4. Of the two carbon substituents, the side chain carbon is bonded to hydrogen, whereas the carboxyl carbon is bonded to oxygen; hence the latter has the higher priority. The overall priority order is, therefore, $N > C$ (carboxyl) $> C$ (side chain) $> H$. When the α-carbon is viewed from the side opposite the hydrogen (priority 4), the decreasing priority of the other groups follows an anticlockwise pattern, and the configuration is designated *S*.

It is a useful exercise to draw out the structure of cysteine; assign priorities to the substituents around the α-carbon according to the C-I-P rules; and use these priorities to assign the configuration of L-cysteine as *R*.

All chiral amino acids found in natural proteins have the *S* configuration, except cysteine where the presence of a sulphur atom only one carbon removed from the chiral centre changes the priority rules about the α-carbon. Thus although cysteine has essentially the same shape as serine, it ends up with the opposite stereochemical descriptor! This kind of anomaly does not deter organic chemists who can see the benefits of the C-I-P rules in a wider context, but is sufficient to make many biochemists prefer the original Fischer terminology. Hence the biological chemistry literature includes both notations and students should be familiar with both.

2.5 Significance of the configurations of α-amino acids

It is not clear whether L-rather than D-amino acids are found in proteins for any reason other than a historical accident propagated by evolution. However, it is intrinsically simpler for nature to use just one form as a building block for proteins. Indeed, for a protein with a defined sequence of 100 amino acid residues in which amino acids are incorporated with either D or L stereochemistry randomly at each position, there are 2^{100} (ca 10^{30}) different stereoisomeric structures. If the stereochemistry is fixed, however, there is only a single stereoisomer for a given sequence. Maintaining the similarity in the configurations of the monomer amino acids therefore leads to well defined structures of the corresponding polymers, i.e. proteins. Since biological chemistry involves the interactions of chiral molecules, the chemistry carried out by cells is greatly simplified by using a single chirality. The vast majority of cells have the facility to construct (*biosynthesize*) L-amino acids and polymers derived from them, and also to break them down (in digestion).

2.6 The stereochemistry of peptide bonds

Proteins are, therefore, linear polymers of L-α-amino acids linked via peptide (amide) bonds. The nitrogen of an amide bond adopts a planar trigonal shape because a partial double bond is formed with the adjacent carbonyl carbon (see Fig. 2.6). This extra bonding is worth about 80 kJ mol^{-1}. The overall stereochemical requirements of a peptide bond are that the carbonyl carbon, the nitrogen, oxygen, hydrogen atoms and the two neighbouring α-carbons are all constrained to lie in a plane. Disruption of this coplanarity by rotation about the carbon–nitrogen bond would lead to loss of some double-bond character. Under biological conditions, this effectively restricts the peptide bond to one of two possible structures: *cis* or *trans*.

Preference for one of these two forms, the *trans* form, arises from a tendency to minimize unfavourable interactions due to too close a proximity of bulky groups (minimization of 'steric crowding' between the substituents on the carbonyl carbon and the nitrogen). Of the two substituents on the carbonyl carbon, it is the α-carbon, rather than the oxygen, that is bulkier. Likewise, the α-carbon is the bigger of the two substituents attached to the nitrogen. For 19 of the 20 protein amino acids, the other substituent on nitrogen is hydrogen; the other amino acid, proline, is discussed below.

The steric preferences of proline are rather different from all the other amino acids since it has an alkyl chain rather than a hydrogen atom as the second substituent on nitrogen. For all amide bonds, except those which involve the nitrogen of proline, there is a significant thermodynamic preference (ca 8 kJ mol^{-1}) for the α-carbon substituents to be as far apart as possible, i.e. for the peptide to exist in a *trans* form.

The D-enantiomers of some amino acids are found in rare instances in nature, often in an effort to avoid the normal metabolism of their L-configured counterparts. An example is D-alanine which is a constituent of the cell walls of bacteria. Peptides formed from this amino acid are fundamentally different in shape from polymers made of L-amino acids. They cannot be degraded by the usual enzymes which digest proteins. This use of D-amino acids can be seen as a protective measure for bacteria to allow them to resist destruction by other organisms. The metabolism of different organisms varies, and this variation can be exploited in the preparation of pharmaceutical agents, as will be noted later. In the case of D-alanine, since its metabolism is essential to bacteria but not humans, molecules which interfere selectively with these metabolic processes are antibacterial agents. The most successful group of antibiotics, the β-lactams (which include penicillins and cephalosporins), owe their activity to the inhibition of enzymes, acting on D-amino acid derivatives, essential to bacterial cell wall biosynthesis (see Section 5.10).

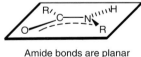

Amide bonds are planar

It is useful to make molecular models of *cis*- and *trans*-dipeptides to convince yourself of the extra steric constraints associated with the *cis* arrangement.

Irrespective of whether a peptide bond to proline is *cis* or *trans*, the α-carbon of the other amino acid is involved in unfavourable steric interactions with a carbon substituent attached to the proline nitrogen.

In natural proteins about 10 per cent of all peptide bonds involving the nitrogen of proline are *cis*. This poses particular problems for generating unique structures for proteins with prolines.

Trans form — Greater steric clash than for a 'normal' *trans*-peptide

Steric clash similar to that of a 'normal' *cis*-peptide bond and comparable to that of the *trans*-peptide bond to proline — Cis form

In the case of peptides involving the nitrogen of proline, this preference is smaller since the α-carbon of the neighbouring residue must avoid either the α-carbon or the other carbon attached to the nitrogen; hence the *cis* form is not dramatically disadvantaged relative to the *trans* form.

2.7 Summary

Proteins are linear polymers of α-amino acids linked via peptide bonds. Each of the 20 monomers bears a characteristic side chain that introduces chemical diversity into the polymer. Side chain substituents include a range of non-polar and polar groups. An appreciation of the acid–base properties of the functional groups found in these side chains is essential to a proper understanding of protein chemistry. The absolute orientation in space of the amino, carboxyl, hydrogen and side chain is the same for all protein amino acids and is usually referred to as the L-configuration, although the *R, S* nomenclature of organic chemistry is also widely used. The peptide bonds that link monomeric amino acids are planar and generally exist in a *trans* rather than a *cis* form. Proline, with its cyclic structure, and glycine, with two hydrogens on the α-carbon, have unusual structural features that set them apart from the remainder of the amino acids.

Further reading

M. Hornby and J. Peach (1993) *Foundations of Organic Chemistry*, Oxford Chemistry Primer, Oxford University Press, Oxford, provides a useful introduction to organic chemical conventions.

J. Clayden, N. Greeves, S. Warren and P. Wothers (2000) *Organic Chemistry*, Oxford University Press, Oxford, and C. K. Mathews, K. E. van Holde and K. G. Ahern (2000) *Biochemistry*, 3rd edn, Benjamin/Cummings, San Francisco, are excellent general textbooks for organic chemistry and biochemistry, respectively.

For information on other aspects of amino acid and protein chemistry, see: J. H. Jones (1992) *Amino Acid and Peptide Synthesis*, Oxford Chemistry Primer, Oxford University Press, Oxford.

3 The structures of proteins

3.1 Overview

The properties of a protein are determined by its shape and its chemical functionality. Proteins form well-defined three-dimensional structures which are the basis of their biological function. The role of some proteins as structural materials can be understood in terms of their molecular architecture; specific examples, including keratins and silk, will be used to illustrate this. Most proteins, however, adopt compact globular structures and this type of structure will be illustrated by that of an enzyme, triose phosphate isomerase.

The overall structure of a protein is related to the order of the monomer units in the chain (*primary structure*) and the conformation that each monomer residue adopts. Where successive monomers adopt the same conformation, regular structural units result. Two classes of structural motif predominate: the α-helix and the β-sheet; these motifs satisfy local steric preferences and optimize longer range interactions, especially hydrogen bonding. These local structural components (*secondary structure*) can be arranged in different ways relative to one another in the overall organization of a protein chain (the *tertiary structure*). Finally, protein chains sometimes associate with each other, and the way they are arranged together is described as the *quaternary structure*. The factors which lead to the formation of specific tertiary and quaternary structures are not fully understood, but a major influence is the nature of the interactions between a protein and the surrounding solvent.

Keratins form the protective covering of all land vertebrates, including materials such as hair, wool, fur, hooves, horns, scales, beaks and feathers.

3.2 Primary structure of proteins

Proteins fulfil a wide variety of roles within organisms. For example, there are thought to be ca 30,000 different proteins in a human body. In any individual protein, the order of individual residues is fixed, imparting specific properties to the final polymer. In recognition of this, a protein is described by its *sequence* (known as the *primary structure*) indicating the order of the amino acid residues in the chain.

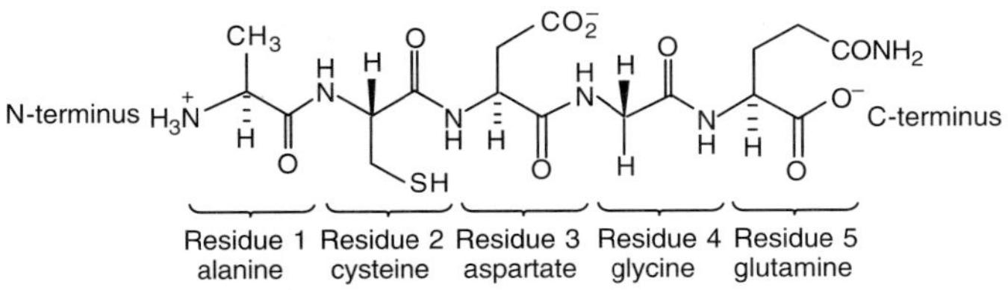

This is: alanyl-cysteinyl-aspartyl-glycyl-glutamine, or H-Ala-Cys-Asp-Gly-Gln-OH
(H and OH are the groups attached to the terminal NH and CO of the peptide chain).

Fig. 3.1 A structure of a representative peptide.

Sequence information can be determined by chemical degradation. Sanger developed a method of reacting the N-terminal residue of a protein selectively and won his first Nobel prize for using this methodology to sequence insulin. This method has been extended by Edman and others and it can readily be automated. This chemistry can, however, only provide information about approximately 20 residues in a single experiment. Sanger won a second Nobel prize for devising a method of sequencing DNA. This methodology can provide huge amounts of sequence information and is the basis of the 'human genome project' which has succeeded in defining the entire sequence of human DNA. Nowadays most protein sequence information is generated by sequencing DNA and inferring the sequence of a protein from the DNA sequence (see Chapter 9).

Three-letter abbreviations are commonly used for protein amino acids. For 16 of the 20, the abbreviation is simply the first three letters of the name: **Ala**nine, **Arg**inine, **Asp**artate **Cys**teine, **Glu**tamate, **Gly**cine, **His**tidine, **Leu**cine, **Lys**ine, **Met**hionine, **Phe**nylalanine, **Pro**line, **Ser**ine, **Thr**eonine, **Tyr**osine-and **Val**ine.

The remaining four amino acids are not so named to avoid ambiguities. The abbreviations for these are: **Asn** for asparagine and **Gln** for glutamine (in each case there is a related amino acid with the same first three letters); **Ile** for isoleucine (which avoids Iso which is a common prefix in chemistry); and **Trp** for tryptophan (which avoids Try which might easily be confused with Tyr).

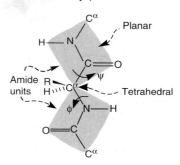

Rotation is only possible about two bonds of the main chain, corresponding to relative rotation of adjacent amide units. The two relevant dihedral angles, ϕ and ψ, are labelled in the diagram.

It is useful to make molecular models of a simple dipeptide and examine, for yourself, the effect of changing the two dihedral angles ϕ and ψ.

Viewing from either end of the bond yields the same value for the dihedral angle—check this for yourself. In the diagram here, the dihedral angles are assessed by viewing from the direction nearest the amino end of the chain, i.e. N to C^α for ϕ and C^α to C for ψ.

The terminology used to describe this primary structure is illustrated for a representative peptide in Fig. 3.1. The ends of the chain are labelled the amino and carboxy, or N and C, termini which in this case are alanine and glutamine residues, respectively. The order of residues in the chain is then given by numbering the monomers from the N to C direction.

3.3 Secondary structure of proteins

3.3.1 Conformational preferences for amino acid residues within a protein chain

An appreciation of the three-dimensional structures of proteins can best be achieved by considering first the stereochemical preferences of a short stretch of a polypeptide chain (secondary structure); structures of complete proteins (tertiary and quaternary structure) are examined in Sections 3.4–3.6.

As discussed in Chapter 2, the α-carbon of each chiral amino acid has the L-configuration and, in a protein, is flanked by two amide linkages both fixed in a planar (generally *trans*) arrangement. This arrangement represents the main chain of a protein. Rotation is only possible around two bonds per residue: the N–C^α and C^α–C bonds; these are described by two dihedral angles ϕ and ψ, respectively (Fig. 3.2). A dihedral angle of 0° corresponds to the situation where the backbone substituents are eclipsed (lie over one another). Positive dihedral angles (up to 180°) correspond to rotating the bond such that the rear substituents move in a clockwise fashion. Conversely, rotation of the rear substituent counterclockwise corresponds to a negative dihedral angle.

The steric effects associated with rotation about these bonds are illustrated in Figs 3.3 and 3.4 for a typical L-amino acid. The conformation of the N–C^α bond where the carbonyl groups lie over one another ($\phi = 0$) involves an unfavourable interaction between these groups. Rotation of the rear substituents of the N–C^α bond in an anticlockwise direction relieves this steric clash. By contrast, an analogous rotation in a clockwise direction relieves this interaction, but only at the expense of introducing a new unfavourable interaction between the side chain and the carbonyl of the preceding residue. Therefore, L-amino acids are expected to prefer conformations where ϕ lies between *approximately* −60° and −180°.

The relative orientation of the bonds of the main chain about the N–C^α bond is the dihedral angle, ϕ.

The relative orientation of the bonds of the main chain about the C^α–C bond is the dihedral angle, ψ.

Fig. 3.2 The dihedral angles ϕ and ψ.

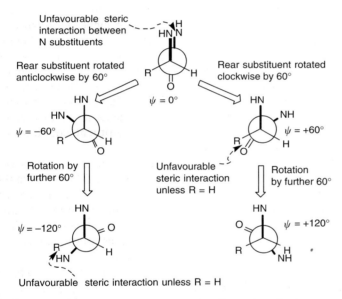

Fig. 3.3 The effect of varying the dihedral angle ϕ.

Note that glycine, which lacks a chiral centre, is less constrained than the other amino acids found in proteins. By contrast, proline is more constrained; it is forced, by its ring structure, to adopt a conformation with a ϕ value of ca $-60°$.

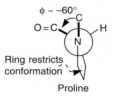

Proline

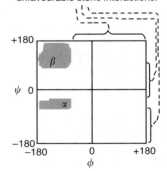

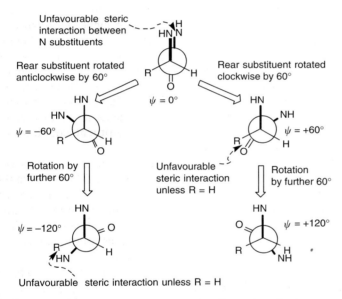

Fig. 3.4 The effect of varying the dihedral angle ψ.

The dihedral angle preferences are conveniently represented by a plot of ϕ values against ψ values, a 'Ramachandran plot', named after the scientist who pioneered this type of analysis.

Fig. 3.5 Stylized Ramachandran plot for an L-amino acid.

A similar analysis of the effects of rotation about the C^α–C bond indicates that L-amino acids will prefer conformations in which ψ is either in the region of $-60°$ or in the region of $+120$ to $+180°$.

In Fig. 3.5, α and β denote regions with favourable dihedral angles that are found in α-helices and β-sheets, respectively.

3.3.2 Regular structures

Given that individual amino acid residues in a protein chain have relatively well-defined conformational preferences, it is reasonable to ask what happens when successive residues adopt the same conformation. Repetition of a specific pair of dihedral angles always gives rise to some form of a helix.

The predicted dihedral angle preferences accord well with the observed values of these angles determined from the structures of proteins as shown in Fig. 3.6.

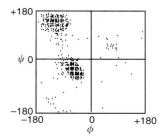

Fig. 3.6 Ramachandran diagram in which the experimental ϕ, ψ angles for a range of residues other than glycine are shown for a representative set of proteins.

The presence of an extra alkyl group rather than a hydrogen atom on the amide nitrogen of proline not only precludes formation of a stabilizing hydrogen bond with the usual carbonyl group but also introduces a steric repulsion with that group. Proline often acts to terminate α-helices and is sometimes called a 'helix breaker'.

The right-handedness of the α-helix is a consequence of the chirality of the α-carbon.

It is remarkable that for an α-helix not only is each of the ϕ, ψ angles such as to minimize steric strain, but all hydrogen bonds are at an optimum length and angle, and all main chain atoms are close packed to maximize van der Waals forces of attraction.

Of the various possible helices, two fit well with the simple steric preferences outlined in Section 3.3.1. These give rise to the most important regular structures, the *α-helix* and the *β-sheet*, which are common components of the structures of proteins (the β-sheet consists of individual β-strands stacked up side by side; each strand is actually a type of helix, see below).

The α-helix ($\phi = -60°$; $\psi = -50°$, Fig. 3.7) is a rather compact helix where, as the chain turns back on itself, linear hydrogen bonds are formed between a carbonyl oxygen and the hydrogen atom of an amide group four residues further down the chain. The overall structure of this unit is a right-handed helix with 3.6 amino acid residues per turn. For every turn of the helix, the chain extends by 0.54 nm (this is termed the *pitch* of the helix). The side chains of the amino acid residues involved in an α-helix point out from the helix and interact with either solvent or neighbouring portions of the protein. All amino acid residues except proline can be readily accommodated within an α-helix (although some have a greater tendency than others to adopt this structure).

The second major class of regular repeated secondary structure found in proteins is the β-sheet, comprised of β-strands ($\phi = -120°$; $\psi = +120°$, Fig. 3.8). A β-strand is an almost completely extended helix; the main chain amide groups, therefore, cannot hydrogen bond with neighbouring residues. They are, however, ideally placed to interact with a neighbouring chain of residues having a similar secondary structure. Two side-by-side

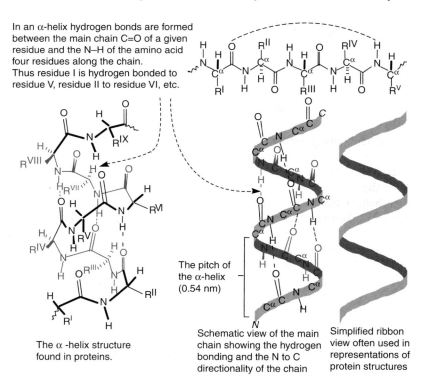

In an α-helix hydrogen bonds are formed between the main chain C=O of a given residue and the N–H of the amino acid four residues along the chain.
Thus residue I is hydrogen bonded to residue V, residue II to residue VI, etc.

The α-helix structure found in proteins.

The pitch of the α-helix (0.54 nm)

Schematic view of the main chain showing the hydrogen bonding and the N to C directionality of the chain

Simplified ribbon view often used in representations of protein structures

Typical amino acid

H-bond donor

No hydrogen on nitrogen. Proline cannot act as a H-bond donor.

Proline

Fig. 3.7 α-Helix.

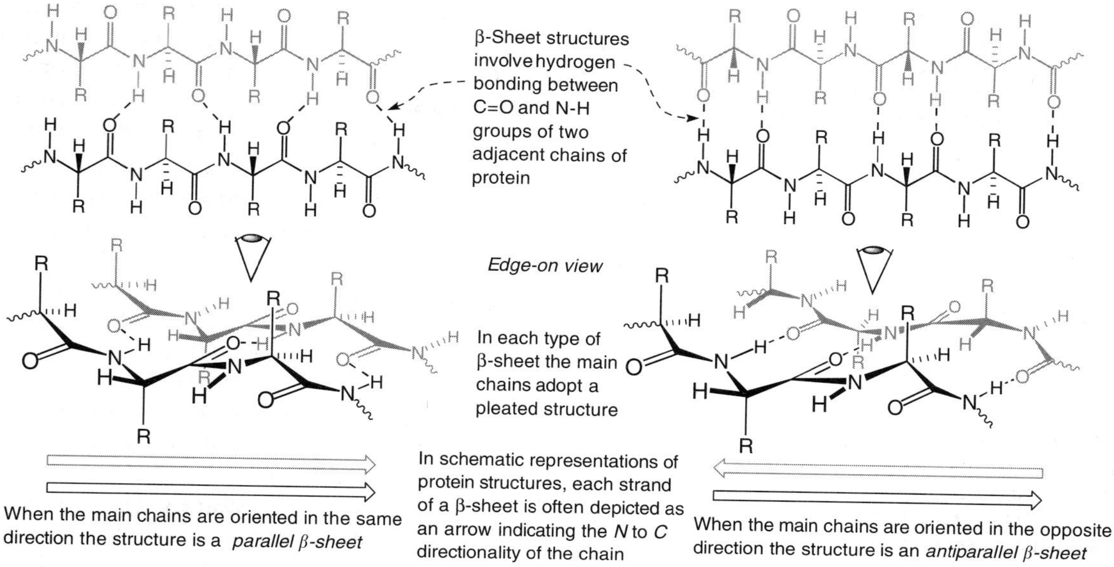

β-Sheet structures involve hydrogen bonding between C=O and N-H groups of two adjacent chains of protein

Edge-on view

In each type of β-sheet the main chains adopt a pleated structure

In schematic representations of protein structures, each strand of a β-sheet is often depicted as an arrow indicating the *N* to *C* directionality of the chain

When the main chains are oriented in the same direction the structure is a *parallel β-sheet*

When the main chains are oriented in the opposite direction the structure is an *antiparallel β-sheet*

Fig. 3.8 The two types of β-sheet structures.

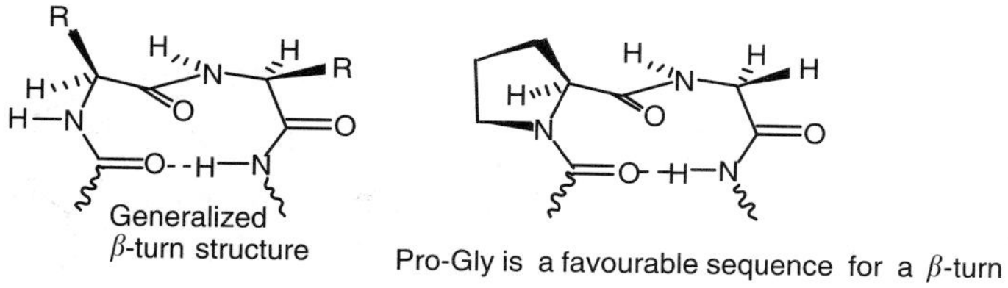

Generalized β-turn structure

Pro-Gly is a favourable sequence for a β-turn

Fig. 3.9 β-Turn structures.

Glycine and proline are often found in β-turns because of their unusual conformational preferences. The structure of proline encourages a turn and the small steric bulk of glycine accommodates the unusual structure.

arrangements are possible in which neighbouring strands are oriented in the same or opposite directions (*parallel* and *antiparallel* β-sheets, respectively). The hydrogen bonding arrangements are somewhat different for these two structures, but in either case a sheet-like structure with a gentle helical twist results. The side chains of individual residues point alternately above and below the plane of the sheet and interact with either solvent or neighbouring portions of the protein.

The hydrogen bond geometry is more favourable in an antiparallel sheet than in a parallel one. The former is more common in proteins.

Connection of the individual strands making up a β-sheet takes the form of a loop. This need not be a regular structure, in the sense that adjacent residues in a turn often adopt different conformations. One class of local structure, the *β-turn*, is found in many protein structures and provides a common means for a protein chain to fold back on itself (Fig. 3.9).

3.4 Tertiary and quaternary structure: α-Keratin

With a knowledge of the secondary structures found in short stretches of protein, the overall structures adopted by proteins can now be analysed.

The reason that the heptad repeat sequence of α-keratin (Fig. 3.10) adopts an α-helical structure is partly because the residues involved have a high propensity for the α-region of conformational space, and partly because of the favourable overall structure which results from packing the helices together.

a	b	c	d	e	f	g
269 Ile	Glu	Ser	272 Leu	Asn	Glu	275 Glu
276 Leu	Ala	Tyr	279 Leu	Lys	Lys	282 Asn
283 Leu	Glu	Glu	Glu	Met	Arg	289 Asp
290 Leu	Gln	Asn	Val	Ser	Thr	296 Gly
297 Leu	Gln	Asn	Val	Ser	Thr	303 Gly

Large non-polar residues are boxed; the numbers refer to the order of the residues in the linear sequence.

Fig. 3.10 A portion of the sequence of mouse α-keratin illustrating the heptad repeat.

It is a useful exercise to draw out the precise structure of each side chain to confirm the polar and non-polar character of the residues in the above sequence.

In the α-keratin structure (Fig. 3.11) the helices are oriented at an angle of about 20° to one another. This type of coiled-coil structure had been predicted by Francis Crick in 1953, before the detailed structure of any proteins were known.

For proteins that reside in an aqueous environment, the burial of non-polar regions is a major driving force favouring the adoption of well-defined overall structures (see Section 1.5). This burial can be achieved by interaction with another non-polar region of the same protein chain to produce an ordered *tertiary structure*. A similar effect can be achieved by the association of separate protein chains to form a cohesive multimeric structure, known as a *quaternary structure*. Specific examples of both types of structure are described below.

The α-keratins, which are key structural proteins in animals, provide the basis for such materials as hair, wool and fingernails. Early X-ray diffraction experiments on proteins included measurements on α-keratins and revealed that these proteins incorporate some regular structural features. These were recognized, after a series of model-building experiments by Linus Pauling, as being due to the presence of α-helices.

When the order of amino acids in α-keratin was determined, regularities were found in the sequence of the main body of the protein. In almost all of this region, chemically similar but not identical residues occur in a repeating cycle of seven residues. It turns out that this sequence, which can be abbreviated as $(a-b-c-d-e-f-g)_n$, predisposes the protein to adopt an α-helical structure; large non-polar amino acids appear at positions a and d and the remaining sites are mostly occupied by polar residues (Fig. 3.10).

Because there are approximately 3.6 residues in each turn of an α-helix, the hydrophobic amino acids are oriented as a ribbon along the edge of the helix (Fig. 3.11). Whilst the remainder of the helix is well solvated by water, this ribbon is not, and there is a thermodynamic driving force to bury this region away from solvent. The extended rod-like structure of α-keratin does not allow this burial to take place within a single protein chain; instead, two chains associate and form a coiled coil. The resulting rope-like structure equips this protein for its role as a constituent of biological fibres.

This discussion has identified several forces that impact on the overall structure of proteins: steric interactions; hydrogen bonding in secondary structural units; and the burial of non-polar residues to avoid unfavourable solvation effects. In addition, the interactions between side chains in tertiary and quaternary structures can be consolidated by covalent cross-linking. As we have seen before, the side chain of cysteine includes a thiol group. If two cysteine residues are sufficiently close in space, a covalent disulphide link can be formed by oxidation. This stabilizes the structure in the form in which the cross-link is made (see Fig. 2.7; although the two cysteine residues must be close in space for disulphide bond formation to occur, they need not be near one another in the protein sequence).

As an example, α-keratin contains some cysteine residues. In hair, these are generally in the form of disulphide links with neighbouring protein chains. These hold adjacent fibres in fixed orientations with respect to one another. Reduction of these links, followed by reorientation of the fibres and reoxidation, changes the overall shape of the hair. This is exploited commercially by hairdressers—it is called a 'permanent wave' or 'perm'.

As was noted in Section 1.4, entropy effects are important in determining the interaction of non-polar residues with water. Burial of non-polar residues, away from water, decreases the unfavourable ordering of the solvent (overall decrease in entropy) that would occur if the non-polar residues were exposed.

When proteins adopt compact folded structures in water, the increased ordering (unfavourable entropy) of the protein chain is more than offset by the increased disorder of water when the non-polar side chains are buried.

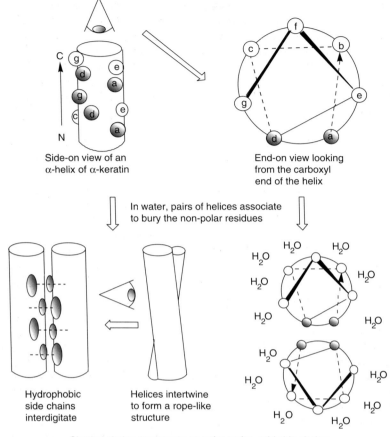

In water, pairs of helices associate to bury the non-polar residues

Shaded circles represent non-polar amino acid side chains;
open circles represent other side chains;
a, b, c, d, e, f and g refer to the order of residues in the
heptad repeat as shown in Fig. 3.10

Fig. 3.11 The structure and packing of α-helices of α-keratin.

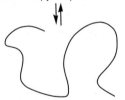

Folded, ordered, protein structure (low entropy for protein chain)

Unfolded, disordered state (high entropy for protein chain)

Disulphide bonds do not significantly affect the degree of order of the solvent or the folded structure, but they restrict the freedom of the unfolded protein chain. Thus, entropically, by comparison with the non-disulphide case, the folded structure is stabilized relative to the unfolded state.

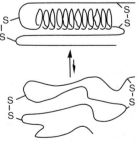

Disulphide bonds reduce the degree of disorder in the unfolded state

3.5 Other structural proteins

There is a large family of proteins that adopt the coiled-coil type of structure found in α-keratin. For example, tropomyosin is an important fibrous protein in muscle. This protein, which contains nearly 300 amino acid residues, also displays a characteristic heptad repeat. Although the structure of this protein is not known in detail, it is closely related (homologous) to that of α-keratin. When proteins have evolved from a common ancestor, they share similar structural features (residues with similar or identical side chains appear in the same order in the related proteins), and they are said to be homologous. This will be illustrated in Chapter 4. The identification of homologies to proteins of known structure is an important aid in the prediction of the three-dimensional structure of a protein from its amino acid sequence.

Although it is possible to rationalize why a protein adopts a particular structure, the prediction of the three-dimensional structure of a protein from its amino acid sequence is extremely difficult. This is because of the complexity of proteins and the folding process by which a particular structure emerges. It is, however, a very important area of research because of the vastly increased knowledge of protein sequences generated by the human genome project and related research. We do not know the structures adopted by the vast majority of these sequences. 'Structural genomics' is a term coined to describe research aimed at generating structural information about proteins which have been identified by genome sequencing.

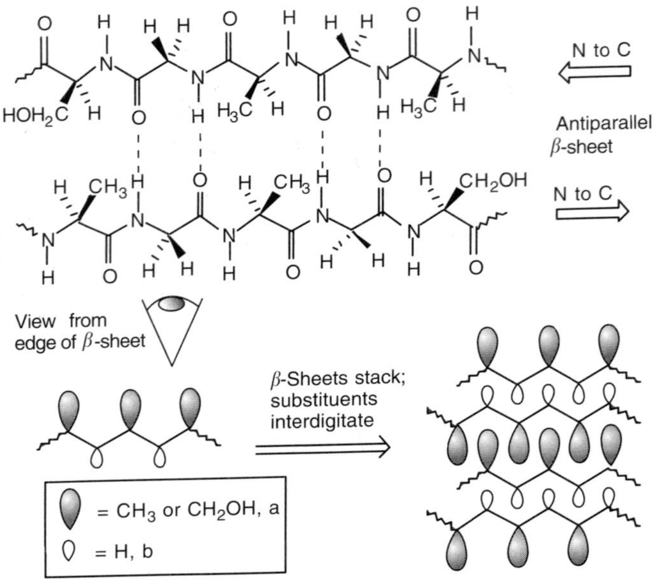

Fig. 3.12 The structure of fibroin.

The crystalline β-sheet region of cocoon silk of the silkworm *Bombyx mori* is responsible for its mechanical strength. This region comprises repeats of the sequence: (Gly-Ala-Gly-Ala-Gly-Ser)$_n$. This forms a rigid stacked structure as shown in Fig. 3.12.

Other regions of this silk protein (fibroin) are glycine rich and adopt an amorphous structure that provides the flexibility of this material.

Collagen is a rigid, triple helix, rich in glycine and proline. The three individual chains intertwine in a coiled coil structure, in a somewhat analogous fashion to that described for the double helix of α-keratin in Section 3.4.

Stacked antiparallel β-sheets are found as key structural features of fibroin proteins, which are found in silk fibres, and of β-keratins, which are components of bird feathers. In a β-sheet structure, the side chains of amino acids point alternately above and below the plane of the sheet. It is not surprising, therefore, that these proteins which form stable repeating β-sheet structures show a repeated diad pattern (a–b)$_n$ (Fig. 3.12), corresponding to the alternation of two types of amino acid residue, over much of their length.

As discussed above, proline cannot be accommodated within a regular α-helix, but proline-rich proteins can form an alternative (extended) helix. Interwound helices of this type are found in collagen, a high tensile strength material which is the principal constituent of connective tissue in animals, including tendons, cartilage and blood vessels.

In conclusion, three basic structures are found in fibrous proteins: the α-helix, the β-sheet and the collagen triple helix. Diverse materials are made by adapting and varying these basic forms. Flexible and elastic materials are often made from α-helical structures with hydrogen bonding within, rather than between, chains. Materials based on β-sheets can be strong and flexible; these structures are exploited particularly by insects as fibres, resistant to stretching. Collagen, a triple-helical structure, is used by animals to make strong, rigid materials, capable of efficiently transmitting mechanical force.

3.6 Globular proteins

Most proteins adopt 'globular' structures in which α-helix and/or β-strand units are linked by turns, allowing the protein chain to fold into a

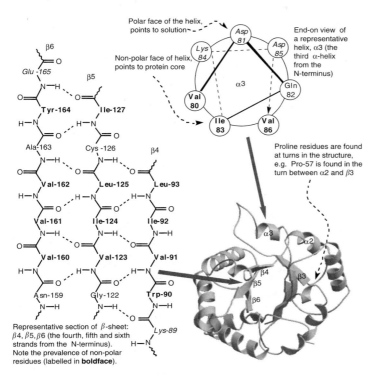

Fig. 3.13 The tertiary structure of triose phosphate isomerase.

The globin family of proteins provides another example of tertiary and quaternary globular protein structures. The oxygen storage protein myoglobin consists of a series of α-helices folded into a compact tertiary structure. In the oxygen transport protein haemoglobin four protein chains, similar to myoglobin, associate to form a well-defined quaternary structure. This is discussed in detail in Section 4.4.

In Fig. 3.13, amino acid residues are numbered according to their order in the linear sequence of TIM. In addition, the polarity of the individual residues is indicated by the way the residues are labelled. Large, non-polar, hydrophobic residues are labelled in **boldface** and charged highly polar residues are labelled in *italics*.

The residues in the middle of the β-strands are non-polar. There are occasional residues in the β-sheet which are polar, but these are at the ends of strands and point into solution, rather than into the core of the protein. As an example, Glu-165, bearing a carboxylate side chain, is at the end of a β-strand. As discussed in Chapter 5, this residue is at the active site of the enzyme; it is on the surface of the protein and interacts with the enzyme substrate. Another key residue at the active site is His-95; this is two residues beyond the end of one of the other β-strands shown. This illustrates how the folded nature of globular proteins brings residues, which are far apart in the primary sequence, close together in space in the final structure.

'Amphiphilic' is derived from Greek, meaning 'lover of both'. An amphiphilic helix is one that likes both polar and non-polar environments, by virtue of having polar and non-polar faces.

The close-packed nature of globular proteins maximizes van der Waals interactions that help to stabilize the structure and provide the rigidity essential for their function.

compact overall structure. In these structures, large non-polar residues are buried in the core of the protein away from solvent water, whereas polar residues are predominantly on the surface of the protein. This type of arrangement is illustrated in Fig. 3.13 for triose phosphate isomerase, TIM, an enzyme that is discussed in detail in Chapter 5.

The three-dimensional structure of TIM has been determined in detail by X-ray crystallography. The enzyme is a symmetrical dimer. The tertiary structure of each TIM monomer (see Fig. 3.13) comprises alternate β-strands and α-helices linked by turns. The overall shape is like a barrel. The eight β-strands at the centre are arranged in a parallel β-sheet, resembling a cylinder. Since this forms the core of the protein, all the amino acid residues in the middle of the β-sheet are hydrophobic. Surrounding the β-sheet are a series of amphiphilic helices, which have a non-polar face pointing towards the protein core and a polar face pointing into solution. The overall structure has a close-packed non-polar core and a predominantly polar surface that interacts with solution. Residual non-polar regions of the surface of individual TIM molecules are buried when they associate to form a dimer, the native quaternary structure of the enzyme. As will be discussed in Chapter 5, the interaction of a particular part of the surface of each protein chain, the 'active site', with substrate molecules is responsible for the catalytic activity of this enzyme.

In order to achieve its functional state after synthesis on the ribosome (see Section 9.11), a protein must fold to the specific structure unique to its particular sequence. The manner in which this occurs is the subject of intense investigation by experimentalists and theoreticians. It is a key link between sequence and structure, and hence to our ability to predict structures, understand protein functions, and design new proteins for specific tasks.

In conditions such as Alzheimer's disease and 'mad cow disease', proteins that are normally globular and soluble misfold and adopt a largely β-sheet structure (related to the fibroin structure described in Section 3.6). In this form the proteins aggregate to form an extended β-sheet structure. The resulting insoluble material that is deposited in the diseased tissue is thought to be at least partly responsible for the characteristic degeneration of the brain associated with these diseases. It is believed that incorrectly folded proteins can promote the misfolding of other molecules, and hence that such proteins, 'prions', can cause some diseases of this type to be transmissible or 'infectious'.

Other diseases associated with protein misfolding include cystic fibrosis and one type of diabetes.

3.7 Summary

Steric interactions limit the conformational opportunities open to most amino acid residues in a protein chain. Two regular structural units, the α-helix and β-sheet, which conform with these requirements, and which exploit favourable hydrogen bonding arrangements, are common in proteins. These secondary structural units, when extended along a sequence, can form fibres. In most proteins, they are connected by loops which allow the protein chain to fold into a compact 'globular' structure. Individual protein chains often fold into regular structures, and sometimes aggregate, in order to bury non-polar regions of the protein that would otherwise be exposed to solvent. The resulting well-defined structures can be consolidated by covalent cross-linking, e.g. by disulphide formation. We now know a great deal about the structures of some proteins and this knowledge provides a solid basis for understanding their properties; however, whilst it is possible to rationalize why a particular structure is preferred, it is difficult to predict the three-dimensional structure of a protein from its amino acid sequence. The generation of three-dimensional structural information for proteins whose sequences are known continues to be a major research challenge, particularly in attempts to define structures and functions for the large number of proteins identified in the sequencing of the human genome. It forms the core of a contemporary field of research known as structural genomics.

Further reading

C. Branden and J. Tooze (1999) *Introduction to Protein Structure*, 2nd edn, Garland Publishing Inc., New York and London, is an excellent introductory textbook in this area.

C. Cohen and D. A. D. Parry (1986) *Trends in Biochemical Sciences*, **11**, pp. 245–8, gives a good general overview of the structure of proteins related to α-keratin.

M. Perutz (1992) *Protein Structure*, W. H. Freeman and Co., New York, provides an overview of protein structure and medicinal applications.

A. R. Fersht (1999) *Structure and Mechanism in Protein Science*, W. H. Freeman and Co., New York, is an outstanding text that discusses many aspects of protein structure and folding.

C. M. Dobson (1999) *Trends in Biochemical Sciences*, **24**, pp. 329–32, gives a brief account of diseases related to protein misfolding.

R. H. Pain (ed.) (2000) *Fundamentals of Protein Folding*, Oxford University Press, Oxford, provides a comprehensive account of protein folding and the diseases associated with misfolding.

4 From structure to metabolic function: the globins

4.1 Overview

In Chapter 2 the basic principles of protein structure were outlined. In Chapter 3 these principles were used to analyse the structures of proteins, including some which serve a purely mechanical function. In this chapter, the role of proteins in metabolism is introduced, using the globins as an example.

Globins are responsible for oxygen transport in many multicellular organisms: oxygen is collected from the atmosphere and delivered to the cells. The molecular mechanism by which this process takes place is one of the best understood examples of the relationship between a structure of a protein and its function, and is examined in detail here.

After a brief introduction, the structures of two such proteins, myoglobin and haemoglobin, will be explored. These provide specific examples of how the basic principles of protein structure can be used to understand the complex functions of large macromolecules. The molecular properties of myoglobin and haemoglobin are then examined in relation to their physiological roles. This is followed by an analysis of the relationship between the structure of the proteins and their function, in atomic detail. Finally, a brief look will be taken at how changes in protein structures affect their function, with sickle-cell anaemia given as an example of a so-called 'molecular disease'.

4.2 Introduction

Many organisms oxidize organic compounds as a means of generating energy (see Chapter 7). Aerobic organisms use oxygen to mediate this chemistry. In unicellular organisms, sufficient oxygen for respiration is available through simple diffusion into cells. In large multicellular organisms the diffusion rate is too slow to fuel the metabolic rate, and a need arises for an oxygen transport and storage system. The globin molecules, a family of proteins, have evolved to fill this need in vertebrates. These proteins bind and release oxygen in response to the needs of the organism; they provide a simple introduction to the interactions between proteins and small metabolites.

The function of haemoglobin is to take up oxygen from the atmosphere and transport it to the various tissues of the body. The function of myoglobin is to collect oxygen from the haemoglobin in the tissues, and store it for later use, freeing the haemoglobin molecules to collect more oxygen. These different, but related, functions are performed by different, but related, proteins.

Oxidation is defined as the loss of electrons. Oxygen is an oxidizing agent because it readily accepts electrons. When an oxygen molecule receives electrons it is reduced, ultimately to water.

Note that myoglobin and haemoglobin are NOT enzymes. Unlike enzymes, the globins do not catalyse chemical transformations of the molecules which they bind. Enzymes are the subject of Chapter 5.

The various globin molecules are similar in sequence and structure; they are 'homologous' proteins. They have evolved from a common ancestor by gradual changes in sequence, to generate a family of proteins with properties tailored to their current biochemical roles. Comparisons of one protein of the family with another provides useful information about structure and function, as well as insights into the evolutionary history of the organisms in which they are found.

A particularly rich source of myoglobin are the muscles of deep-sea diving mammals, such as whales. Why?

The behaviour of myoglobin (Mb) can be understood in terms of the equilibrium between oxygenated and deoxygenated forms:

$$MbO_2 \rightleftharpoons Mb + O_2$$

The equilibrium constant for the dissociation of oxymyoglobin, K_{Mb}, can be written as

$$K_{Mb} = [Mb][O_2]/[MbO_2]$$

A fraction, y_{Mb}, of the myoglobin molecules have bound oxygen (Fig. 4.1); the remainder $(1 - y_{Mb})$ are deoxygenated. Hence,

$$K_{Mb} = (1 - y_{Mb})[O_2]/y_{Mb}$$

which can be rearranged into the form

$$y_{Mb} = \frac{[O_2]}{K_{Mb} + [O_2]}$$

This equation describes the behaviour shown in Fig. 4.1. At high concentrations of O_2, $[O_2] \gg K_{Mb}$ and y_{Mb} is approximately $[O_2]/[O_2]$, i.e. it approaches 1. At low concentrations of O_2, $[O_2] \ll K_{Mb}$ and y_{Mb} approximates to $[O_2]/K_{Mb}$, i.e. it is proportional to the O_2 concentration. K_{Mb} is a measure of the affinity of myoglobin for O_2; a small value of K_{Mb} corresponds to a high affinity for O_2. A similar mathematical relationship is found in Chapter 5 describing the properties of proteins which act as catalysts.

Myoglobin is a monomer. Haemoglobin is a tetramer in which each of its four component subunits resembles myoglobin. The subunits of haemoglobin, however, have somewhat different properties from myoglobin both because of small differences in primary structure and by virtue of the interaction of the four subunits. The differing behaviour of myoglobin and haemoglobin provides a useful lesson in how subtle changes to protein structure are used to fine-tune important aspects of their function.

4.3 The physiological role of the globins

Myoglobin is an oxygen storage protein and is particularly abundant in skeletal and cardiac muscle; it binds oxygen at all but the very lowest concentrations encountered in active muscle. Oxygen is only released when the concentration of oxygen has fallen to very low levels. Haemoglobin is the oxygen transport protein and is abundant in the blood; it binds oxygen in the lungs and releases it in the capillaries where oxygen is required by respiring tissues.

In order for the globins to fulfil their respective roles, haemoglobin must have a lower affinity for oxygen than myoglobin at all oxygen concentrations. This feature is clearly illustrated in Fig. 4.1. It is also apparent from the graph that the oxygen affinities of myoglobin and haemoglobin vary with oxygen concentration in a different fashion. It is important to recognize the physiological significance of this difference in oxygen binding patterns, as is also shown in Fig. 4.1, and to explain how this difference is achieved at a molecular level.

The behaviour of myoglobin can be readily explained by assuming that each molecule has a constant binding affinity for oxygen irrespective of the oxygen concentration. At very low concentrations of oxygen, most myoglobin molecules do not contain a bound oxygen molecule. As the concentration of oxygen increases, an increasing proportion of myoglobin

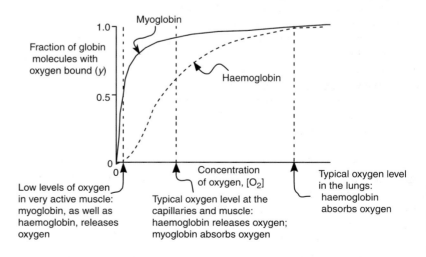

Fig. 4.1 The oxygen binding properties of myoglobin and haemoglobin.

molecules have oxygen bound; they are now referred to as *oxymyoglobin*. At higher oxygen concentrations, the solution of myoglobin becomes saturated as all the myoglobin is in this oxy form.

The behaviour of haemoglobin, however, is clearly more complicated than that of myoglobin and cannot be explained in terms of the simple arguments above. As the oxygen concentration increases, the affinity of the remaining haemoglobin molecules for oxygen appears to increase. This phenomenon, known as *cooperativity*, puzzled biochemists for many years.

The cooperative binding of oxygen to haemoglobin allows it to carry out its biological role very efficiently. At low oxygen concentrations, most of the oxygen is released to the tissues, whereas at high oxygen concentrations (e.g. in the lungs) virtually all molecules of haemoglobin become saturated with oxygen.

The structures of myoglobin and haemoglobin, with and without bound oxygen, have been elucidated and are examined in the next section. The differences between the structures are then used to account for their different behaviour in Section 4.5.

4.4 The structure of the globin molecules

4.4.1 The structure of myoglobin and oxymyoglobin

Myoglobin was the first globular protein to have its structure determined in atomic detail. Many of its features are now well understood, and have proved common to a large number of protein structures that have since been elucidated. For example, the residues in myoglobin are found to be very efficiently packed together (see Fig. 4.2); also, the hydrophobic amino acid residues are found to be buried in the core of the protein, whilst the more hydrophilic residues are exposed at the surface, as discussed in Section 3.6.

The protein chain of myoglobin (Fig. 4.2) contains 153 amino acid residues and is made up of 8 secondary structural units, all α-helices. In contrast to the extended α-helical structure found in α-keratin (Section 3.4), the α-helices in myoglobin are connected by short loop regions that allow the chain to fold back on itself to make a compact overall structure. The lengths of the helices range from 7 to 28 amino acid residues. Careful inspection of the tertiary structure reveals that helices adjacent to one another in the protein sequence are not always close to one another in space.

Detailed examination of the full structure of myoglobin allows us to rationalize the arrangement of helices using some of the principles of secondary structure which were outlined in Chapters 2 and 3. For instance, the α-helices often pack against one another at an angle of approximately 50°; this angle allows a favourable interaction between a ribbon of side chains protruding from the surface of one helix and the groove bordered by two such ribbons in the other helix. This is one arrangement of α-helices that occurs commonly in protein structures.

The eight helices (A–H) enclose a hydrophobic pocket in which resides a haem group. This pocket is the active site of a globin molecule and it binds a single oxygen molecule; it will be examined in more detail in Section 4.5.

The cooperative behaviour of haemoglobin provides a classic example of *allosteric* behaviour in proteins. Allostery, the phenomenon whereby the binding of a molecule at one site on a protein surface alters the functional properties of the protein at a distant site, is an important mechanism for regulating biological phenomena. Cells control a wide variety of processes by regulating the activity of key proteins. This regulation is brought about by controlling the amount of these proteins (see the discussion of the regulation of gene expression in Section 9.11) and their activity within the cell. Allostery plays a key role in the latter type of regulation.

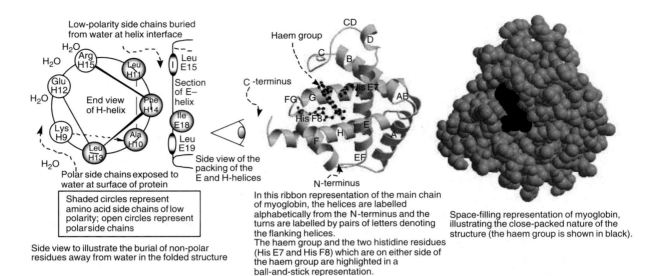

Fig. 4.2 The structure of myoglobin.

In this discussion of the globins, the amino acid residues are identified by their position relative to the N-terminal end of each individual helix, e.g. His F8 refers to a histidine residue which occurs as the eighth residue within the F-helix.

The haem group is an organic molecule, not a protein (see Fig. 4.5). Such a molecule that is permanently associated with a protein, in its functional state, is termed a prosthetic group. These species are closely related in function to coenzymes (see Chapter 7) and are used when the intrinsic chemistry of the normal amino acid side chains is insufficient to perform the chemical function required of the protein.

From a structural viewpoint it is important to note that both the hydrophobic nature and the precise geometry of the haem pocket, so critical to its function, are preserved in all known globins.

Oxymyoglobin has almost the same structure as myoglobin: the oxygen molecule is bound to the haem group causing a slight change in tertiary structure, specifically at the F-helix.

4.4.2 The structure of haemoglobin and oxyhaemoglobin

All vertebrate haemoglobins have been found to have similar structures: they comprise a tetramer of globin molecules, each molecule containing a haem group in a hydrophobic pocket. Each haem group can bind one oxygen molecule, which means that each haemoglobin molecule may carry a maximum of four molecules of oxygen. The protein chains are generally of two types, termed the α and β chains, which have *slightly* different sequences and tertiary structures. Most haemoglobins are made up of two copies of each type of monomer and can be represented as $(\alpha\beta)_2$.

Each protein chain has a structure closely resembling that of myoglobin, as can be seen in Fig. 4.3. The four chains are arranged symmetrically in a tetrahedral array. This arrangement creates a roughly spherical molecule, with the haem groups separated from one another in space. The arrangement of the four protein chains within the tetrameric assembly is termed the quaternary structure of haemoglobin. The interactions between the four chains are responsible for the allosteric properties of haemoglobin and will therefore be examined in some detail. Most of these interactions are between α and β chains, as will be explained below.

It turns out that the origin of the allosteric behaviour of haemoglobin results from the fact that this protein can exist in two slightly different forms,

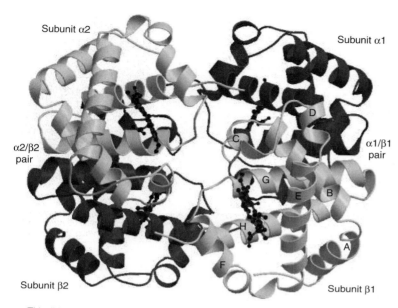

This ribbon representation of haemoglobin shows the main chains of the four sub-units α1, α2, β1 and β2. The right- and left- hand halves, α1/β1 and α2/β2 act as pairs. The haem groups are highlighted as black ball-and-stick representations.

Fig. 4.3 The structure of haemoglobin.

known as *R* and *T* states. At high oxygen concentrations, haemoglobin is predominantly oxygenated and in the *R* form, which has an affinity for oxygen comparable to the free myoglobin-like subunits; at low oxygen concentrations, haemoglobin is substantially deoxygenated and most molecules are in the *T* form, which has a considerably lower affinity for oxygen than free subunits. The differences in structure between the two forms must account for the difference in oxygen affinities. The challenge for the chemical biologist is to rationalize this relationship at an atomic level.

Detailed studies have shown that the structures of haemoglobin in its *R* and *T* forms differ primarily in the relative orientation of the individual protein subunits. A close inspection of the structures of the two forms of haemoglobin shows that, in addition to minor changes in tertiary structure, specific αβ-subunit interactions (Fig. 4.4) differ; there are more salt bridges and hydrogen bonds in the *T* form of haemoglobin than in the *R* form. This extra stabilization of the *T* form encourages the adoption of this low oxygen affinity structure in the absence of oxygen. The binding of oxygen offsets this stabilization, favouring the *R* form, as will be described in section 4.5.

On transition from the *R* to the *T* form, interactions between subunits α1 and β1 and between α2 and β2 tend to remain constant; indeed haemoglobin can be regarded as a dimer of αβ dimers. However, the interactions between subunits α1 and β2 and α2 and β1 change; as such they are known as sliding contacts. Interconversion of the *R* and *T* forms can

The problem of predicting the full structure of the globins from their amino acid sequence alone should be apparent from Fig. 4.3. The importance and difficulty of predicting protein structures from their sequences was discussed in Chapter 3.

In Fig. 4.3 one of the monomers, β1, is drawn in a similar orientation to that of myoglobin in Fig. 4.2 and the helices are labelled A–H to allow a structural comparison to be made.

By comparison with the analysis of oxygen binding by myoglobin in Section 4.3, the behaviour of haemoglobin, Hb (Fig. 4.1), can be described by an equation of the form

$$y_{Hb} = \frac{[O_2]^n}{K_{Hb} + [O_2]^n}$$

where y_{Hb} is the fraction of haemoglobin molecules with oxygen bound and K_{Hb} is an effective equilibrium dissociation constant of oxyhaemoglobin. This equation reflects the fact that haemoglobin can bind up to four molecules of O_2. If the binding of oxygen molecules by each chain were completely independent, then haemoglobin would have similar oxygen binding properties to myoglobin, i.e. $n = 1$. On the other hand if the binding of oxygen were an 'all-or-nothing' process, the equilibrium would depend on $[O_2]^4$, i.e. $n = 4$. In practice, the behaviour of haemoglobin is in between these two extremes and can be described by the above equation where $n = 2.8$.

'Salt bridge' is a general term for the electrostatic attraction between adjacent side chains bearing complementary charges. This is an important stabilizing interaction for charged residues of proteins that are not fully exposed to solvent.

During oxygen binding to haemoglobin, the change from the *T* to the *R* form involves a net reduction in the number of salt bridges and hydrogen bonds, and hence involves an expenditure of energy. This explains why the affinity of haemoglobin for oxygen is always lower than that of myoglobin–an extra energy price has to be paid when O_2 is bound.

As shown in Fig. 4.4, the $\alpha\beta$ contact acts as a two-way switch.

Note how a detailed understanding of the structural changes requires a knowledge of the specific chemical functionality of the particular amino acids involved. It is useful to refer back to Chapter 2 to remind yourself of the precise molecular structures of the amino acids concerned.

In the *T* to *R* transition one $\alpha\beta$ dimer turns relative to the other

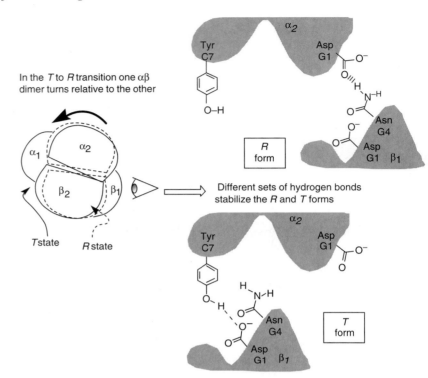

Different sets of hydrogen bonds stabilize the *R* and *T* forms

Fig. 4.4 Schematic diagram showing structural differences between the *T* and *R* forms of haemoglobin. (Redrawn after Perutz (1990).)

be described in global terms as a 15° rotation of one $\alpha\beta$ dimer relative to the other (Fig. 4.4); this change in structure is the basis of the mechanism of cooperativity of haemoglobin which is discussed in the next section.

4.5 The molecular mechanism of oxygen transport

The haem group of myoglobin imparts the characteristic red colour to muscle tissue, whilst haemoglobin is responsible for the red colour of blood.

The role of His F8 provides another example of how the specific character of an amino acid side chain is exploited in a functional protein structure.

The haem group consists of an Fe^{2+} ion encased in a porphyrin ligand. The Fe^{2+} ion has six coordination sites; the porphyrin ligand takes up four of these, leaving two free binding sites on opposite sides of the metal ion. If a free haem group is present in aqueous solution, these sites may be occupied by water molecules. In globin molecules, one site is coordinated to a histidine residue (His F8) and the other is available to bind oxygen, being protected from solvent molecules by the protein chain (Fig. 4.5).

In an aqueous solution containing oxygen, a free Fe^{2+}-haem group binds oxygen molecules but, unlike the globin molecules, binding results in irreversible oxidation of the Fe^{2+} ions to Fe^{3+}. The oxidation destroys the oxygen-binding activity of the iron.

In globin molecules, the oxygen molecules are bound to the haem reversibly, and no oxidation occurs. The protein environment protects the Fe^{2+} from oxidation. The exact mechanism of protection has been investigated by studying the active site of the globins, and then synthesizing

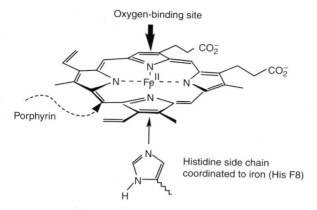

Fig. 4.5 The environment of Fe^{2+} in the haem group of the globins.

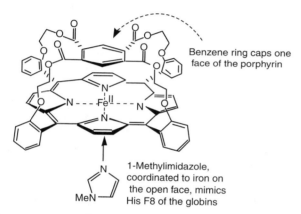

Fig. 4.6 A synthetic mimic of myoglobin: a porphyrin modified to allow reversible binding of oxygen.

Simple iron porphyrins irreversibly oxidize, in the presence of oxygen, via a series of dimerization processes as outlined below:

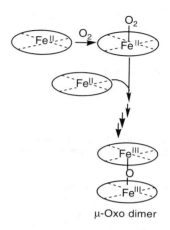

The capped porphyrin shown in Fig. 4.6, designed by Baldwin, can bind oxygen reversibly. When oxygen binds to this porphyrin, it cannot interact freely with solvent or other porphyrins as outlined above. Instead the cap protects the oxygen-binding site as shown below:

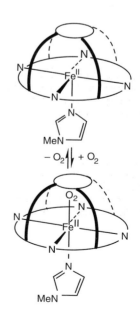

porphyrins which mimic this environment, thereby establishing the requirements for reversible oxygen binding. In globins, the haem with bound oxygen is somewhat buried in a hydrophobic environment. Access to solvent is restricted, and there is no opportunity for the haem to interact directly with other haem groups.

Synthetic porphyrins (Fig. 4.6) have been prepared which have a hydrophobic 'cap' over the oxygen-binding site of the iron. Oxygen bound at this site cannot interact directly with solvent, or with other porphyrins. Such molecules do indeed bind oxygen reversibly. These studies are important for two reasons: they help identify exactly the features required for an oxygen-carrying protein, and they may provide the basis for the development of artificial blood.

4.5.1 The nature of the oxygen-binding site

The geometry of the haem group in myoglobin is depicted in Fig. 4.7. When oxygen binds to the Fe^{2+}, it causes a change to its electronic

The synthetic porphyrins, which reversibly bind oxygen, provide a good example of biomimetic chemistry. In this field of research, chemistry is used to provide insights into biological systems. In turn, such research provides new compounds with interesting properties that can sometimes be exploited in technological applications. For example, compounds that reversibly bind oxygen have potential as components of artificial blood.

The electronic structure of an Fe^{2+} ion is $1s^2\ 2s^2\ 2p^6\ 3s^2\ 3p^6\ 3d^6$; there are six electrons in the outer d shell. These electrons are organized in one of two distinct ways. In one configuration the d electrons are present as three electron pairs. Since there is no residual electron spin, this is known as the low-spin configuration. Such ions have a relatively small ionic radius. The alternative is the high-spin arrangement in which the maximum number of electrons are unpaired. In this form the Fe^{2+} ion adopts an electronic configuration with two of the d electrons paired and the other four unpaired. Such ions are larger than the low-spin variant. Different ligand environments around the Fe^{2+} ion favour different configurations. In the absence of oxygen, the Fe^{2+} ion adopts the high-spin configuration. The binding of oxygen favours a change to the smaller, low-spin, form as shown in Fig. 4.7.

In its deoxy form, the haem porphyrin distorts from the normal planar array to a domed structure, in an effort to accommodate the large Fe^{2+} ion, as shown schematically in Fig. 4.7.

structure which results in a slight reduction in its atomic radius. This allows the Fe^{2+} to move into the plane of the porphyrin ring, which in turn pulls His F8 towards the haem group. This process explains the small difference in tertiary structure observed between myoglobin and oxymyoglobin (Fig. 4.7).

Although the change in the electronic configuration of the Fe^{2+} on oxygen binding causes only a small change in tertiary structure, this transition has been exploited as the crucial trigger associated with the transition from the *R* form to the *T* form in haemoglobin. The movement of His F8 necessarily results in a movement of the F-helix of which it is a part. Since this helix is a rigid entity, it acts as a lever and magnifies the small movement, transmitting a larger change to the subunit interface. Thus, movement of the F-helix favours the rotation of the $\alpha\beta$ dimers relative to one another and, in so doing, forces the salt bridges (which stabilize the *T* form) to break.

Progressive binding of oxygen to haemoglobin, therefore, increases the stability of the *R* state, relative to the *T* state, hence increasing the relative population of the high-affinity form of the haemoglobin molecules in solution. In other words, the binding of oxygen to one subunit of haemoglobin has facilitated a change in the structure of the tetramer, which increases the affinity of the remaining haem groups in the molecule for oxygen. In this way, communication between subunits is achieved, and the phenomenon of cooperativity of oxygen binding to haemoglobin is

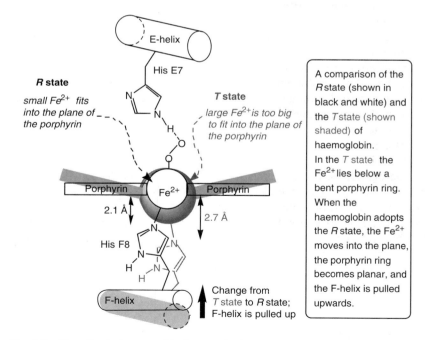

A comparison of the *R* state (shown in black and white) and the *T* state (shown shaded) of haemoglobin. In the *T* state the Fe^{2+} lies below a bent porphyrin ring. When the haemoglobin adopts the *R* state, the Fe^{2+} moves into the plane, the porphyrin ring becomes planar, and the F-helix is pulled upwards.

Fig. 4.7   The effect of oxygen binding on the haem group and adjacent parts of the structure of haemoglobin. (Redrawn after Perutz (1990).)

explained. This phenomenon is now understood in great detail. Indeed, it has proved possible to set up a detailed mathematical model and simulate the oxygen-binding curves shown in Fig. 4.1 very closely.

4.6 How changing protein structure alters protein function: sickle-cell anaemia and molecular disease

Sickle-cell anaemia provides the classic example of a disease which can be explained in terms of an unwelcome change in protein structure; as such, it has been termed a 'molecular disease'.

The symptoms of sickle-cell anaemia include headache, weakness and dizziness, all of which are attributable to a lack of oxygen. These symptoms are particularly acute in situations where the body's oxygen supply is restricted, e.g. at high altitudes or during vigorous exercise. An examination of the patient's red blood cells reveals that many are sickle shaped, rupture easily and cause capillary blockage.

More detailed investigation reveals that this sickling is a consequence of a change in the haemoglobin itself, which has an abnormally low solubility in its T state. At low oxygen concentrations, where the T state predominates, the defective haemoglobin tends to form a fibrous precipitate which deforms the cells and gives them their distinctive sickle shape.

The cause of this low solubility turns out to be a single amino acid substitution in a key position on the surface of the defective haemoglobin: a valine has been substituted for a glutamate at position 6 in the β chain. This apparently small change in structure has a profound effect on the properties of the haemoglobin molecule. This is because valine is a hydrophobic residue and, unlike the polar glutamate, prefers to be buried in the interior of the protein. A single haemoglobin molecule is unable to bury the valine residue within its hydrophobic core. Aggregation of the defective deoxyhaemoglobin molecules, however, happens to be a favourable process, due to the presence of a complementary hydrophobic area in the EF corner of each β chain in neighbouring molecules. This area is not exposed to the surface in oxyhaemoglobin, which explains why the problem is exacerbated at low oxygen concentrations.

Aggregation of defective deoxyhaemoglobin molecules results in long helical fibres of protein molecules which can be readily observed in an electron micrograph. The single amino acid substitution results in a small local change in structure, but changes the overall protein structure from a soluble, globular tetramer, to an extended fibre, and disables the oxygen transport function. An understanding of the principles of protein structure and function has allowed us to explain this clinical phenomenon in molecular terms.

4.7 Summary

Myoglobin is an oxygen storage molecule prevalent in muscle tissue. Haemoglobin is an oxygen transport molecule found in red blood cells.

Chemicals that bind to the T structure, and stabilize it, disfavour the T to R transition. This will, therefore, diminish the affinity of haemoglobin for oxygen. This effect is exploited as a means of regulating the oxygen binding of haemoglobin. As an example, the pH of muscle is lower than that of the lungs. Protons bind to, and selectively stabilize, the T state. This property, known as the 'Bohr effect', enhances the efficiency of release of oxygen by haemoglobin in the tissues. Molecules which affect the allosteric equilibrium are known as 'allosteric effectors'. Other examples of species which act as allosteric effectors for haemoglobin are Cl^-, CO_2 and a metabolite 2,3-bisphosphoglycerate. The latter is involved in the adaptation of humans to high altitude: changes in its concentration act to enhance the efficiency with which oxygen is offloaded to the tissues.

Sickle-cell anaemia is more prevalent amongst certain ethnic groups, e.g. people of central African descent. The chronic effects of sickling are only manifest when an individual has inherited mutant haemoglobin genes from both parents; humans who have inherited a mutant gene from one parent, and a normal version from the other are relatively unaffected by anaemia. This group, however, is more resistant to the effects of malaria. In countries where malaria is endemic, there is therefore a genetic benefit in having a single copy of the sickle-cell gene. This explains why sickle-cell anaemia has not disappeared in the course of evolution.

Chapter 3 outlined some diseases related to protein misfolding. Sickle-cell anaemia provides another example of a disease caused by the aggregation of proteins into an insoluble fibrous precipitate.

Drawing the precise molecular structures of glutamate and valine is a useful way of understanding the change in chemical functionality that is involved in the sickle-cell mutation.

The two molecules have very different oxygen affinities over a range of oxygen concentrations and this allows them to fulfil their respective biological roles. In particular, haemoglobin is found to bind oxygen cooperatively: the binding of one molecule of oxygen enhances the binding of the next.

Myoglobin has a monomeric structure with eight α-helices that pack to form a hydrophobic pocket. This pocket binds a haem group containing Fe^{2+} in a specific geometry; the haem group is the active site of the molecule and binds the oxygen. Haemoglobin is a tetrameric molecule comprised of four myoglobin-like subunits; it exists in two forms, R and T, which differ in oxygen affinity. The existence of two forms with different affinities explains the cooperative behaviour of haemoglobin, and can be rationalized after examining the difference in the quaternary structure of the two forms, and the tertiary structures of the individual subunits.

Detailed experiments suggest that the molecular mechanism of oxygen transport depends on reversible binding of oxygen to the iron atom in the haem group; the reversibility of this process is facilitated by the specific binding environment. A change in electronic configuration of the iron, on binding oxygen, provides an atomic trigger that accompanies a change in structure from the R to T form.

Small changes in the structure of the protein can have profound effects on its function. Sickle-cell anaemia provides a classic example of this phenomenon: a single amino acid substitution results in defective haemoglobin and clinically detectable symptoms.

Further reading

C. Branden and J. Tooze (1999) *Introduction to Protein Structure*, 2nd edn, Garland Publishing Inc., New York and London, Chapter 3, describes structural details of α-helical proteins including myoglobin.

General biochemistry texts, such as those listed in Chapter 1, describe aspects of the biochemistry of myoglobins and haemoglobin. Another biochemistry text with extensive coverage of the globins is L. Stryer (2000), *Biochemistry*, 4th edn, W. H. Freeman and Co., Basingstoke.

M. F. Perutz (1990) *Mechanisms of Cooperativity and Allosteric Regulation in Proteins*, Cambridge University Press, Cambridge, provides a comprehensive account of allostery in general, including details of the structure and molecular mechanism of haemoglobin.

5 Proteins as catalysts

5.1 Overview

As we have seen, proteins adopt well-defined three-dimensional structures and are capable of binding small molecules reversibly at one or more specific sites on their surfaces. This property is exploited by some types of proteins, *enzymes*, to catalyse chemical reactions. Enzymes generally catalyse specific reactions of particular molecules and the enhancements of reaction rate are often remarkably high; indeed some enzymes are so efficient that the rate-limiting factor in the catalysed reaction is the diffusion of the reactants to the enzyme.

An appreciation of the chemistry of enzyme catalysis provides insights into the biochemical details of cells, as well as into chemical catalysis itself. Enzyme chemistry is also the basis for much important applied science, e.g. the development of novel pharmaceutical compounds. Many features of enzyme catalysis will be illustrated by considering one important enzyme, triose phosphate isomerase, which is understood in great detail. The principles that emerge in considering this enzyme can then be extended to provide a general appreciation of our understanding of enzymes and the methodology by which these fascinating catalysts are studied.

5.2 Triose phosphate isomerase

Cells carry out a vast range of chemical reactions in an organised fashion. The chemical transformations essential to cells are catalysed by enzymes, as required by the organism. One of these reactions, that catalysed by triose phosphate isomerase (TIM), is considered here as an illustrative example. The reaction catalysed by TIM is a key step in the chemistry used by cells to derive chemical energy from the oxidation of sugars (see Chapter 7).

The names of enzymes are often derived by adding the suffix -*ase* to the name of the substrate and the nature of the reaction catalysed. Hence 'triose phosphate isomer*ase*' is an enzyme which isomerizes triose phosphates.

TIM catalyses the interconversion of two phosphorylated sugars: D-glyceraldehyde-3-phosphate (D-GAP) and dihydroxyacetone phosphate (DHAP). Here, the details of this reaction are analysed, and some experiments that have led to our current knowledge of the mechanism of the catalytic process are outlined. The chemical transformation of the aldehyde D-GAP to the ketone DHAP is relatively simple: a carbonyl group and alcohol group are interconverted.

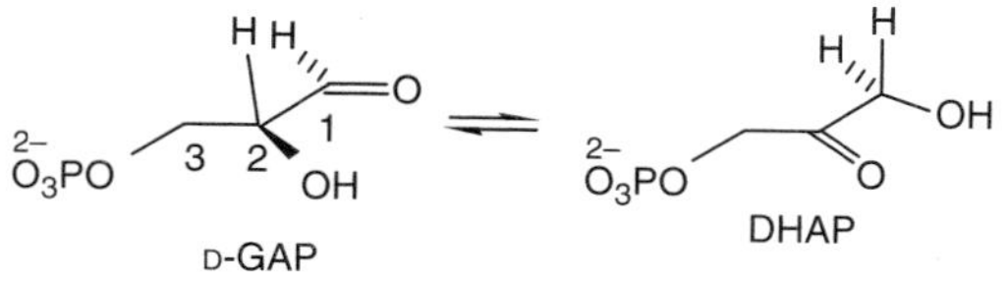

The structures and chemistry of phosphorylated sugars, including D-GAP and DHAP, are discussed in Chapter 6. The numbering scheme used for phosphorylated sugars is described there; it is shown here for D-GAP since it is used later in this chapter.

The interconversion of D-GAP and DHAP can be accelerated somewhat, in the absence of TIM, by simple acid- or base-catalysis.

The isomerization of D-GAP and DHAP is a reversible process.

$$\text{D-GAP} \underset{k_{-1}}{\overset{k_1}{\rightleftharpoons}} \text{DHAP}$$

As the reaction proceeds, D-GAP and DHAP are constantly being interconverted. Irrespective of the relative starting concentrations of D-GAP and DHAP, the reaction reaches a point (the equilibrium position) where the concentrations of the two species reflect their relative thermodynamic stabilities. At this stage, D-GAP and DHAP are being interconverted at equal rates, and no overall change in their relative concentrations results. The ratio of the concentrations at this point is the equilibrium constant, K:

$$K = [\text{DHAP}]/[\text{D-GAP}]$$

The equilibrium constant can also be described in terms of the rate constants for the forward and back reactions, k_1 and k_{-1}, respectively. At equilibrium the rate of formation of DHAP is equal to the rate at which DHAP is reverting to D-GAP, i.e.

$$k_1[\text{D-GAP}] = k_{-1}[\text{DHAP}]$$

Therefore

$$[\text{DHAP}]/[\text{D-GAP}] = k_1/k_{-1}$$

Hence the equilibrium constant, K, is equal to k_1/k_{-1}.

An assay used for the measurement of catalysis by TIM is mentioned in Section 7.4.3.

5.3 The rate of the reaction catalysed by TIM

When a sample of D-GAP is dissolved in water, in the absence of a catalyst, it slowly isomerises (Fig. 5.1). Eventually, a stable mixture results at a point where 96 per cent of the aldehyde has been converted to DHAP. This final composition reflects the equilibrium position of this reaction in aqueous solution.

If the experiment is repeated in the presence of even a small amount of TIM, a more rapid reaction takes place, but the final equilibrium position is the same. TIM acts as a *catalyst* accelerating the rate of attainment, but not the position, of the equilibrium. Detailed inspection of the TIM-catalysed reaction shows that the initial rate of conversion of D-GAP to DHAP (represented by the slope of the curve at time zero (Fig. 5.1)) can be very rapid. This rate gradually slows until, at the equilibrium point, no overall change takes place. Only at the very start of the reaction does the measured reaction rate correspond simply to the conversion of aldehyde to ketone. As ketone accumulates, the reverse reaction (ketone to aldehyde) becomes significant, decreasing the *overall* (net) rate (but *not* the *intrinsic* rate) of conversion of aldehyde to ketone. The measured rate of reaction near equilibrium is therefore difficult to analyse in a simple manner. Hence, most meaningful studies of enzyme catalysis use initial rate data.

Information about the nature of catalysis can be obtained by measuring the effect on the reaction rate of changing the reaction conditions. Firstly, a procedure is required to measure the rate of reaction precisely. Then, keeping all other factors constant, the initial rate of reaction is measured as different variables are altered in turn.

Varying the concentration of TIM changes the rate of the enzyme-catalysed reaction, as shown in Fig. 5.2. The initial rate varies in direct proportion to the concentration of TIM present: if the enzyme concentration is doubled, the rate doubles; trebling the enzyme concentration trebles the rate, and so on. In the terminology of physical chemistry, the reaction is *first order* in enzyme.

The effect of changing the concentration of D-GAP, whilst keeping the amount of enzyme fixed, is illustrated in Fig. 5.3. At low concentrations of substrate, the initial rate changes in direct proportion to the amount of

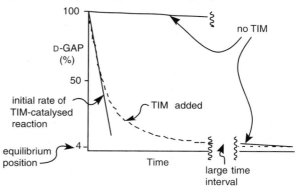

Fig. 5.1 The isomerization of D-GAP to DHAP.

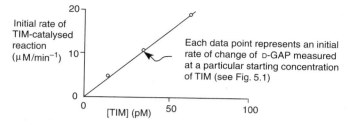

Fig. 5.2 Measured initial rate of reaction vs concentration of TIM.

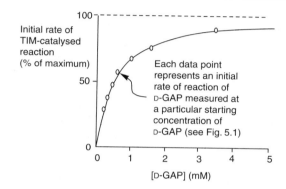

Fig. 5.3 Measured initial rate of reaction vs starting amount of D-GAP.

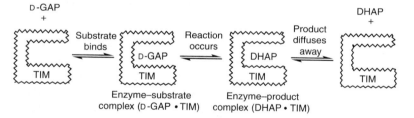

Fig. 5.4 Schematic overview of catalysis by TIM.

The rate behaviour shown in Fig. 5.3 is similar in form to that found for O_2 binding by myoglobin (see Section 4.3). These phenomena can both be understood in terms of the reversible binding of a molecule at an active site on the surface of a protein.

The initial rate of the TIM-catalysed reaction (when there is no significant back reaction) can be represented by a simplified scheme where TIM reversibly binds the substrate, D-GAP, and catalysis takes place on the bound substrate:

$$D\text{-}GAP + TIM \rightleftharpoons D\text{-}GAP \bullet TIM \xrightarrow{k} DHAP + TIM$$

The initial rate of reaction will be

$$rate = k[D\text{-}GAP\bullet TIM]$$

The concentration of enzyme–substrate complex is the total enzyme concentration, [TIM], multiplied by the fraction of TIM molecules, y_{TIM}, with bound substrate:

$$[D\text{-}GAP\bullet TIM] = [TIM]y_{TIM}$$

Substrate binding is analogous to the O_2-binding behaviour of myoglobin (see Section 4.3). Thus it can readily be shown that the fraction of TIM molecules with bound substrate is of the form

$$y_{TIM} = \frac{[D\text{-}GAP]}{K_{TIM} + [D\text{-}GAP]}$$

where K_{TIM} is a constant related to the dissociation equilibrium of substrate from the enzyme.

Hence the rate is given by

$$rate = k[TIM]\left(\frac{[D\text{-}GAP]}{K_{TIM} + [D\text{-}GAP]}\right)$$

substrate. Under these conditions, the reaction is first order in substrate. As the substrate concentration is increased to higher levels, the initial rate gradually approaches a maximum value. The way in which the rate varies as the concentration of D-GAP is changed provides information about the nature of the enzyme-catalysed process.

This type of behaviour can be understood by analogy to the binding of O_2 by myoglobin described in Chapter 4. As we have seen, myoglobin stores oxygen in the muscle by binding the oxygen at a specific site on the surface of the protein. The amount of oxygen molecule bound by myoglobin increases as the concentration of oxygen increases. Eventually, at high concentrations of oxygen, effectively all the binding sites are occupied and the myoglobin is saturated with O_2. This saturation behaviour is analogous to that of an enzyme such as TIM in the presence of its substrate. A simple scheme for catalysis by TIM, drawn by analogy with myoglobin, and annotated for the conversion of D-GAP to DHAP, is shown in Fig. 5.4.

The kinetic behaviour of TIM is typical of many enzymes. This behaviour is generalized in the Michaelis–Menten equation:

rate of formation of product

$$= k[E]\left(\frac{[S]}{K_m + [S]}\right)$$

where E and S represent enzyme and substrate, respectively, k is the catalytic rate constant and K_m is the Michaelis constant.

At high concentrations of substrate, essentially all the enzyme molecules are engaged in catalysis. The fraction of enzyme with substrate bound, $[S]/(K_m + [S])$, approaches 1 $(K_m \ll [S])$ and the Michaelis–Menten equation simplifies to

$$\text{rate} \approx k[E]$$

$k[E]$ is therefore the maximum rate, representing the point at which the enzyme saturates, and increasing the concentration of S no longer influences the rate of product formation.

At low substrate concentrations, $[S] \ll K_m$ and the fraction of enzyme with substrate bound is approximately $[S]/K_m$. The Michaelis–Menten equation then simplifies to

$$\text{rate} \approx \frac{k[E][S]}{K_m}$$

i.e. the rate is first order in enzyme and substrate. The ratio of constants, k/K_m, is the effective rate constant under these conditions.

Insights into the mechanism of chemical reactions, the nature of the bond-forming and bond-breaking processes, are usually established by postulating different alternatives, and then undertaking experiments which distinguish between them. In this way, incorrect mechanisms can be discounted. A mechanism can never be proved correct, but a mechanism that fits all the information generated in an intensive study is likely to be close to the truth.

In the first stage of catalysis, the substrate binds reversibly at the active site of TIM; the substrate then undergoes a chemical reaction. At low concentrations of substrate, the active sites of relatively few TIM molecules are occupied; molecules of D-GAP can therefore bind and react each time one meets an enzyme molecule. Hence the rate is proportional to the concentration of D-GAP as well as that of TIM. At high concentrations of D-GAP, however, most of the enzyme molecules are already involved in catalysis; some D-GAP molecules therefore encounter TIM molecules with occupied active-sites. Since the substrate cannot bind, these encounters cannot lead to an enzyme-catalysed reaction. The rate is therefore no longer simply proportional to the concentration of D-GAP. This simple scheme explains why the reaction rate reaches a maximum when the active site is saturated at high concentrations of D-GAP.

5.4 The chemical nature of the TIM-catalysed reaction

The catalytic behaviour of TIM has so far been rationalised in terms of a process whereby D-GAP is bound at the active site of the enzyme and, when bound, undergoes a chemical reaction to produce DHAP. To explain the dramatic rate enhancement brought about by TIM, we need to develop an understanding of the molecular details of the chemistry that takes place on the enzyme surface.

Various mechanisms can be postulated for the isomerisation of D-GAP to DHAP, catalysed by TIM. Two chemically reasonable possibilities are shown in Fig. 5.5. In (a), a hydrogen atom migrates within the molecule from C-2 to C-1, whereas in (b), a proton (hydrogen ion) is removed from C-2 and then, in a subsequent step, a proton is added to C-1.

These possibilities have been investigated using a radioactive labelling experiment (Fig. 5.6). An isotopically labelled form of D-GAP was made with radioactive hydrogen (tritium) attached to C-2. When TIM catalysed the conversion of tritium-labelled D-GAP to DHAP, some of the radio-activity appeared in the solvent, rather than DHAP. This is incompatible with mechanism (a), proving that the hydrogen does not exclusively migrate within the molecule.

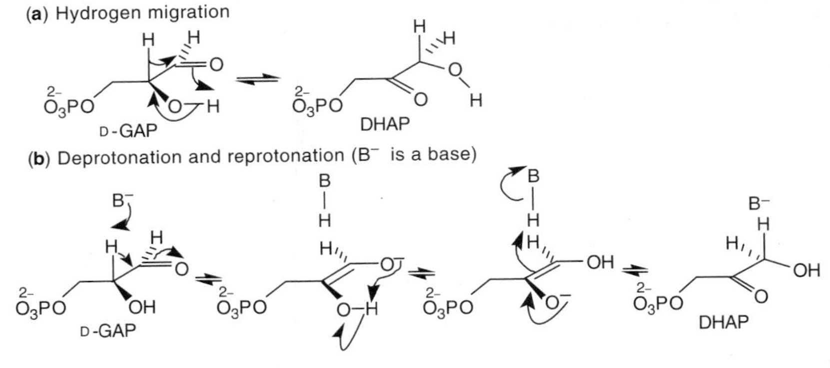

Fig. 5.5 Two mechanistic possibilities for the interconversion of D-GAP and DHAP.

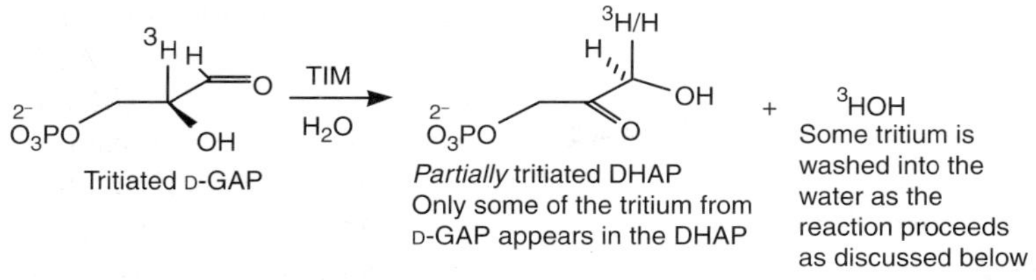

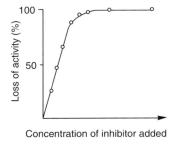

Fig. 5.6 Investigating the mechanism of TIM using tritiated D-GAP.

Mechanism (b) is compatible with the outcome of the tritium-labelled D-GAP experiment. This mechanism implies that a basic functional group is located on the protein surface (at the active site). The next stage in unravelling the mechanism was to investigate whether there is an amino acid residue capable of acting as a base at the active site of TIM.

5.5 Identifying the active site of TIM

Molecules that resemble D-GAP or DHAP, but have different chemical properties, have been used to identify the active site of TIM. One such molecule, which is sufficiently similar in structure to DHAP to bind to the active site of TIM, is iodohydroxyacetone phosphate. As is typical of alkyl iodides, iodohydroxyacetone phosphate undergoes alkylation reactions with nucleophiles (the iodide ion is a good leaving group).

When TIM was incubated with iodohydroxyacetone phosphate, the enzyme was irreversibly inhibited (converted into a form incapable of catalysing the isomerization of triose phosphates) as shown in Fig. 5.7. The degree of inhibition increased as more inhibitor was added, until no enzyme activity remained. At this point, one molecule of inhibitor had become covalently attached to each protein chain. The inhibitor was attached to a particular glutamate residue, Glu-165 (see Section 3.6).

The simplest explanation of these results (see Fig. 5.8) is that Glu-165 is present at the active site of TIM in the correct position to act as a base for the removal of a proton required for isomerisation. When iodohydroxyacetone phosphate is incubated with TIM, it binds to the active site, in the same way as does a normal substrate. In this case, however, the glutamate acts as a nucleophile and forms a covalent bond to this substrate analogue. Once securely attached to the active site of the enzyme, there is

Fig. 5.7 Inhibition of the catalytic activity of TIM as a function of the concentration of iodohydroxyacetone phosphate added.

Tritium was used because its chemistry is analogous to that of normal hydrogen and its radioactivity allows it to be detected readily (only very small amounts of the tritiated molecules need be present). The fate of the tritium in the reaction can be determined and used to infer what happens to the equivalent hydrogen in the 'normal' reaction.

Mechanisms based on deprotonation chemistry are reasonable, since protons attached to carbons adjacent to a carbonyl group are moderately acidic; a typical pK_a value would be approximately 20. Since the side chains of amino acid residues in proteins contain both acidic and basic functional groups, it is also reasonable that the enzyme could use acid- and/or base-catalysis.

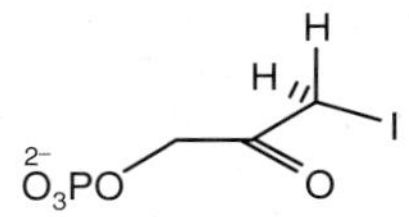

iodohydroxyacetonephosphate

The number of inhibitor molecules attached to each protein molecule, and their site of attachment, were determined by inhibiting TIM with radioactively labelled iodohydroxyacetone phosphate. This compound forms a covalent bond to the enzyme. Measuring the amount of radioactivity of the inhibited enzyme showed that 1 mol of inhibitor reacts with 1 mol of enzyme. The radioactively labelled enzyme, generated by this experiment, was degraded into a mixture of small peptide fragments. Of these, only the fragment that had reacted with the inhibitor was radioactive. This peptide was isolated. Determination of the number and order of the amino acids present allowed its position in the overall primary structure of TIM to be identified. In turn, the radioactivity of this peptide was found to reside at the point corresponding to a single glutamate residue, Glu-165, which was,

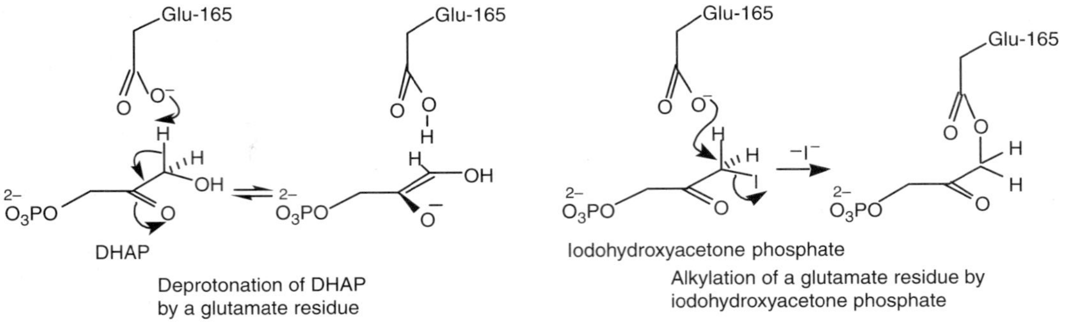

Fig. 5.8 The chemistry of DHAP and iodohydroxyacetone phosphate at the active site of TIM.

therefore, tentatively identified as the active-site base.

Labelling experiments, of the type used to identify Glu-165 as an active-site residue, must be interpreted carefully. Checks must be made that the inhibitor has actually reacted at the active site, rather than elsewhere on the enzyme. Furthermore, the presence of an amino acid capable of acting as a base does not *prove* that that residue acts as a base in the enzyme-catalysed reaction.

During evolution the sequences of proteins alter. Changes of amino acids that are involved in catalysis are more likely to reduce the effectiveness of the enzyme and so are less likely to persist. When the sequences of enzymes from various organisms are compared, a disproportionate number of the conserved amino acids (those which occur at the equivalent place in each linear sequence) are often associated with active sites. A comparison of 13 sequences of TIM, isolated from different sources, reveals that 44 of approximately 250 residues are completely conserved (identical in all sequences) including Glu-165. Two-thirds of these conserved residues are within 12Å of the pocket containing Glu-165, identified as the active site.

no possibility that this molecule can diffuse away to free the active site for catalysis. Since it is no longer possible for a substrate to bind at the active site, the enzyme is inhibited.

The sequence of TIM has been determined for enzymes produced by a wide variety of organisms. The sequences are similar and, in each case, the monomeric protein has a molecular weight of approximately 26,000. The three-dimensional structure of the enzyme has been determined by X-ray crystallography (Fig. 3.13) and was discussed in Section 3.6.

The structure of TIM is complex, but insights into its catalytic role can be obtained without studying all of it in detail. The majority of the protein can be regarded as a support for the active site which has the right shape, chemical functionality and flexibility to carry out the catalytic process. Glu-165, the residue implicated as the catalytic base, lies in a pocket close to one end of the barrel (see p27). Close examination of this region of the surface of the protein reveals other amino acid residues with side chains close to that of Glu-165 and which, therefore, may play a role in the catalytic events. Of the other amino acid residues, lysine at position 13 (Lys-13) and histidine at position 95 (His-95) are best positioned to assist in catalysis. Notice that these amino acid residues are close together in space, but come from very different parts of the protein chain.

In this particular case, confirmatory evidence that this region is the active site has been obtained by determining the X-ray structure of TIM in the presence of DHAP. The DHAP molecule was found to bind in the pocket adjacent to Glu-165 and very close to Lys-13 and His-95. Such detailed information allows firm conclusions to be drawn about the exact nature of catalysis by TIM.

5.6 A possible mechanism for TIM catalysis

A mechanism for the TIM-catalysed interconversion of triose phosphates that accommodates all the experimental information described above is shown in Fig. 5.9. Firstly, D-GAP binds at the active site, with the hydrogen atom on C-2 close to Glu-165. The glutamate removes a proton to form an intermediate anion which can subsequently accept a proton at C-1, to produce the product DHAP. The product can then diffuse out of the active site.

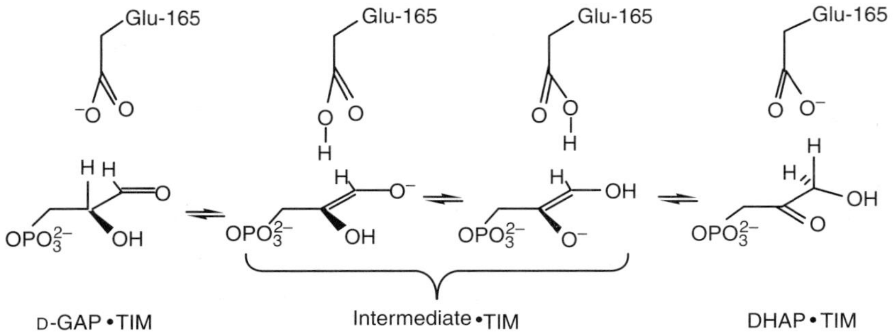

Fig. 5.9 A mechanism for the interconversion of D-GAP and DHAP catalysed by TIM.

Proton transfer between oxygen atoms, such as that shown for the intermediate in Fig. 5.9, occurs readily in water.

In the intermediate, the proton removed from D-GAP becomes attached to Glu-165. This proton can then be exchanged with protons in the solvent. This exchange is the basis for the observed washout of radioactivity from tritiated D-GAP, described in Section 5.4.

The overall reaction in the presence of TIM proceeds a billion (10^9) times faster than the same process in solution catalysed by a simple carboxylate ion, such as acetate. Indeed, the limiting factor in the reaction rate is the speed at which substrates and products can diffuse to and from the enzyme. How can this dramatic rate acceleration be explained in molecular terms?

Several factors can be identified which contribute to the remarkable rate enhancement brought about by TIM. Because of the range of functional groups available on amino acid side chains, enzymes are able to use acid- *and* base-catalysis. In contrast to conventional chemistry, where one might employ either acids (e.g. hydrochloric acid) *or* bases (e.g. sodium hydroxide) to speed up reactions, enzymes can use both types of catalysis simultaneously, and all at pH 7. This is possible because the enzyme surrounds the substrate, and different functional groups interact specifically with different parts of the substrate.

The enzyme binds the substrate in a precise fashion at the active site, prior to any chemistry taking place. The subsequent chemistry is more efficient than non-enzymatic reactions occurring between pairs of molecules in free solution: at the active site there are no solvent molecules to interfere with the chemistry, and everything is preorganized and optimally aligned for reaction.

The nature of the binding of substrates, products and intermediates to enzymes is a distinguishing feature of the observed catalysis. Energy profiles of the acetate-catalysed and TIM-catalysed interconversion of D-GAP and DHAP are illustrated in Fig. 5.10. These diagrams show the relative free energies of the different species that occur as the reaction proceeds. They depict information about the rate of the reaction (the lower the energy barrier, the faster the reaction) and equilibrium position (the relative energies of the starting material and product).

Isomerisation of D-GAP involves passing through a high-energy transition state and producing a deprotonated intermediate. The formation of DHAP from this intermediate occurs in a similar fashion. The acetate-catalysed reaction (Fig. 5.10a) is relatively slow because there is a large energy difference between D-GAP (and DHAP) and the transition states. This energy barrier must be overcome for isomerization to occur. The formation of the intermediate is unfavourable because it is higher in

In the interaction of substrate with the enzyme, the same range of forces that are available to stabilize protein structures (Chapter 3) can be employed to facilitate chemical transformations: electrostatic attraction and repulsion (of charged and dipolar entities), along with hydrogen bonding and weaker interactions such as van der Waals forces (although individual interactions of this type are small, the great number possible when an enzyme envelopes a substrate can produce a significant cumulative effect—a good analogy is with a Velcro fastener, where the cumulative effects of many individual hooks lead to a strong grip).

The energy profile shown in Fig. 5.10 depicts ΔG for the interconversion of D-GAP and DHAP. The energy of D-GAP is little affected on binding to TIM, but DHAP is actually slightly *destabilized*. Raising the energy of a reactant complements the lowering of the energy of the high-energy species: both factors narrow the energy gap which has to be surmounted.

An important feature of enzyme catalysis is that the rate of one reaction is enhanced, whilst side reactions are minimised. The geometry of D-GAP, bound at the active site, is ideal for deprotonation; it also suppresses the loss of phosphate by β-elimination, which is a major competing reaction in the 'normal' chemical isomerization.

When D-GAP binds to TIM, the phosphate group is held away from the plane of the hydrogen that is removed:

A different conformation of D-GAP is required for β-elimination:

β-elimination is facile when the C–H and C–OPO$_3{}^{2-}$ bonds, broken in the reaction, are in the same plane.

By controlling the conformation of the bound D-GAP, TIM controls its fate.

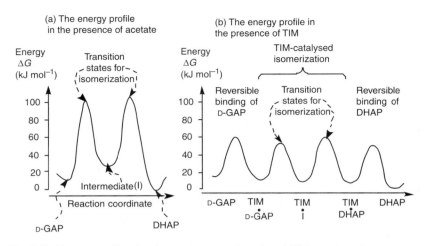

Fig. 5.10 Energy profiles for the acetate-catalysed and TIM-catalysed interconversion of D-GAP and DHAP.

energy than either D-GAP or DHAP. The overall process is, however, favourable since DHAP is lower in energy than D-GAP.

The energy profile of the reaction catalysed by TIM involves two additional steps: the reversible binding of D-GAP and DHAP to TIM (Fig. 5.10b). These binding steps do not greatly affect the energies of D-GAP and DHAP. In the TIM-catalysed reaction, however, the intermediate and the adjacent transition states are greatly stabilized; this can be seen by comparing the two energy profiles (Fig. 5.10a and b). The net effect of lowering the energies of the transition states and the reaction intermediate is roughly to halve the energy barrier to isomerisation. This corresponds to the enormous rate enhancement that is observed in the TIM-catalysed reaction.

In conclusion, detailed experiments have revealed that TIM binds substrates and products relatively weakly, and intermediates and transition states tightly. This selective binding of high-energy species is often termed 'transition state stabilization'. The high-energy species 'fit better' at the active site than do the substrates and products. The net effect is that the energy barriers to reaction have all been reduced to the point where the isomerization proceeds as fast as is physically possible, i.e. as fast as diffusion to, and from, the active site of the enzyme allows.

5.7 Testing the proposed mechanism

Dealing firstly with the enzyme, the mechanistic and structural studies suggest that the functional groups of Glu-165 and His-95 are essential for efficient catalysis. If the nature or position of these groups (carboxylate and imidazole, respectively) is changed, catalytic efficiency should drop. Developments in genetic engineering make it possible to change amino acids of interest within a protein sequence. Changing Glu-165 to an aspartate residue decreases the rate of reaction by a factor of 1000; even a small change in the orientation of the carboxylate group of this crucial amino acid, relative to the substrate, is sufficient to disrupt catalysis. When His-95 is changed to a glutamine, the catalytic effectiveness drops more than 200-fold. This result indicates the importance of the imidazole functional group of histidine.

Turning now to the substrate, the proposed mechanism for catalysis by TIM involves the generation of a high-energy intermediate during the reaction. One source of catalysis is believed to be the tight binding of this species by the enzyme. If this is so, then small molecules which resemble this high-energy intermediate should also bind tightly and compete with substrates for the active site. This will decrease the rate of the reaction.

Several molecules which resemble the proposed high-energy intermediate in the D-GAP–DHAP interconversion have been examined for their effect on catalysis by TIM. For example, phosphoglycolohydroxamate has been found to bind very tightly to TIM and thereby inhibit the reaction. It binds at the active site and the structure of the enzyme with this inhibitor bound has been determined by X-ray crystallography (Fig. 5.11). Structures of this type have provided important information about the enzyme–substrate geometry during catalysis.

5.8 General principles of enzyme catalysis

A relatively detailed analysis of the catalysis of a reaction by one enzyme, TIM, has been presented. The focus on one example has been used to give a flavour of what is known about enzyme catalysis and the experiments which provided these insights. The concepts and methodology that have been highlighted for TIM are applicable to a wide range of enzymes.

Enzymes, for the most part, catalyse reactions that have direct counterparts in organic chemistry. Moreover, the mechanisms of enzyme-catalysed reactions are usually directly related to those of organic chemistry. The principal differences lie in the speed and, especially, the specificity of enzyme-catalysed processes. The origin of the latter difference (the fact that only certain molecules react and these do so only in particular ways) lies in the fact that enzyme catalysis involves the binding of reactants at an active site on the protein. This binding process is specific to

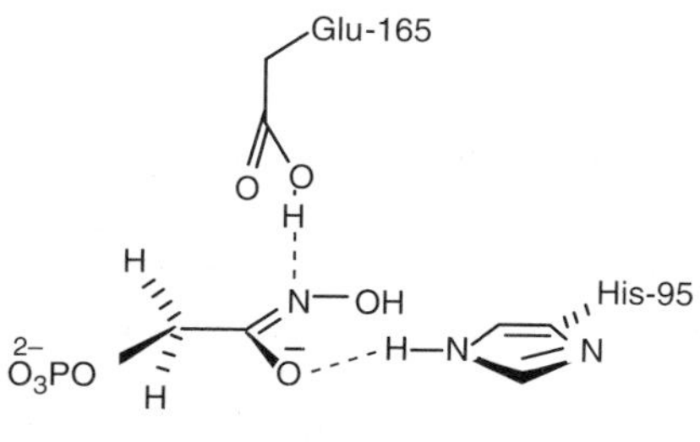

The primary sequence of a protein is determined by genetic information carried in the form of DNA. DNA, and the proteins it encodes, can be manipulated by 'genetic engineering'. The DNA that encodes a protein of interest can be modified using a technique called 'site-directed mutagenesis'. The protein with a change in the sequence can then be produced from the modified DNA.

An introductory account of DNA, and genetic engineering, is given in Section 9.12. In the mutagenesis experiments, glutamate at position 165 was changed to aspartate since the only difference between these residues is a shortened side chain; the carboxylate group is retained. Glutamine was chosen to replace histidine at position 95 since it is similar in size and polarity, but lacks the imidazole group. Here, differences in catalysis can be related to the changed functional group.

X-ray crystallography has provided detailed information about the interactions between TIM and substrates, and related compounds. Glu-165 and His-95 both hydrogen bond to substrates and analogues. They are ideally placed to act as acid–base catalysts. Further experiments have reinforced the view that they act in this fashion in the catalytic process.

Phosphoglycolohydroxamate is a structural analogue of a proposed intermediate in the TIM-catalysed reaction:

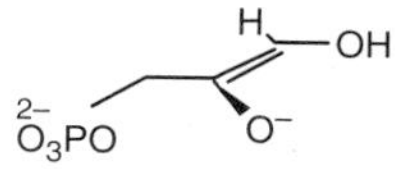

Molecules which resemble a high-energy intermediate of a reaction are often effective enzyme inhibitors. They are frequently, but somewhat misleadingly, labelled collectively as 'transition state analogue inhibitors'.

Fig. 5.11 Schematic view of the active site of TIM with an inhibitor bound.

molecules that have the right shape and functional groups to interact with the enzyme.

Once bound at the active site, the substrate is in contact with several of the amino acid residues of the protein chain. Since these residues may contain acidic, basic and nucleophilic groups, reactions which can be catalysed by these types of functionalities can be brought about by proteins without the need for additional reagents. Reactions that require other types of chemistry, such as oxidative or reductive processes, can also be catalysed by enzymes; in order to do so, however, enzymes conscript other molecules or ions (termed coenzymes) which have the requisite properties (see e.g. Section 7.4).

Enzymes catalysing readily reversible reactions generally bind substrates and products only weakly. They do, however, decrease the barrier to reaction by binding high-energy species tightly. This selective use of binding and the benefits of proximity, controlled reaction geometry and exclusion of interfering solvent molecules leads to many enzymes being extremely efficient catalysts. They can speed the reaction of interest and suppress possible side reactions. Enzymes which are such good catalysts that the rate-limiting factor in the reaction is diffusion of reactants to and from the active site of the enzyme are sometimes called 'perfect' catalysts.

5.9 The methods of enzyme chemistry

Since a primary function of enzymes is to enhance reaction rates, measuring the rates of these reactions gives information about enzyme action. A detailed analysis of how the rate of an enzyme-catalysed reaction is altered by changing the reaction conditions provides great insights into enzyme catalysis.

Determining the structure of the enzyme by X-ray crystallography, once we know the location of the active site, can be used to identify the amino acids involved in catalysis. The roles of these amino acids can be evaluated by changing them, in a controlled fashion, using site-directed mutagenesis.

Isotopically labelled substrates and inhibitors are useful in probing the details of enzyme chemistry and inhibition. Enzyme inhibitors, in general, play a major role in enzyme chemistry. Two important kinds of enzyme inhibitor have been mentioned in this chapter. Firstly, molecules which resemble substrates for an enzyme, but which are electrophilic, can react with nucleophiles at the active site of an enzyme. Once they are covalently tethered at the active site, the enzyme is prevented from carrying out further chemistry. Such inhibitors provide information about the identity of the active site. Secondly, molecules which resemble high-energy intermediates (or transition states) involved in the reaction can bind non-covalently, but tightly, to the enzyme active site. These inhibitors can provide information about the mechanism of catalysis.

In addition to their utility in unravelling the details of enzyme action, enzyme inhibitors have important technological applications. Since

---

The use of coenzymes for catalysis is analogous to the use of iron and a porphyrin group by myoglobin and haemoglobin to bind oxygen, as described in Chapter 4; this process could not be effected by a protein chain alone.

A variety of methods is available to obtain values of the rate parameters (k and K_m) from measured rate data. One simple method (although not the most accurate) is the double reciprocal or 'Lineweaver–Burk' plot, where the reciprocal of the rate is plotted against the reciprocal of the substrate concentration (1/rate vs 1/[S]). The values of the maximum rate and the Michaelis constant can be obtained from the intercepts.

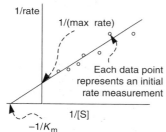

Each data point represents an initial rate measurement

The Michaelis constant, K_m, is the value of [S] at which the initial rate is half its maximum value. It is a measure of the affinity of an enzyme for a substrate: small K_m values indicate that the enzyme has a high affinity for substrate and vice versa.

enzyme-catalysed reactions are essential to the correct functioning of a cell, disruption of an enzyme-catalysed process is usually debilitating. A molecule that inhibits an enzyme catalysing a reaction essential to all cells is likely to damage any cell it enters, and so will probably be very toxic. Knowledge of such toxins is important for human health.

Enzyme inhibitors are not all bad, however. In fact they form a major part of a very important business: the pharmaceutical industry. An analysis of the role of enzyme inhibitors as therapeutic drugs will conclude this discussion.

5.10 Enzyme inhibitors as drugs

The range of reactions carried out by different organisms varies. For instance, it is essential for bacteria to carry out some reactions that have no human counterpart. We can exploit these biochemical differences for therapeutic ends. A compound which inhibits an enzyme which is essential to an organism that humans would like to kill (such as some bacteria or fungi) can be used to destroy the organism. If humans have no equivalent enzyme, then they may remain unaffected by this compound. For example, penicillins (Fig. 5.12), probably the most famous of all antibiotics, inhibit a family of bacterial enzymes which catalyse the production of strong outer cell walls. Since human cells lack walls of this type, we do not have equivalent enzymes. Penicillin therefore kills bacteria, but is generally innocuous to humans. The careful use of penicillin is of great benefit to human society. Since the discovery of penicillin, many other therapeutically useful enzyme inhibitors have been found. The success of antibacterial and antifungal drugs testifies to the importance of enzyme inhibitors, and they will continue to be exploited for many years to come in fighting disease.

Measuring the precise way in which an enzyme-catalysed reaction rate is affected by an inhibitor provides detailed insights into the nature of inhibition.

The use of antibiotics has contributed greatly to human health, especially during the latter half of the twentieth century. Bacteria, however, counter this challenge. They evolve means of becoming resistant to antibiotics. A bacterium that resists the effects of an antibiotic can grow where others cannot; they are at an advantage in environments where antibiotics are present. Resistant strains of bacteria can, therefore, be selected by an inappropriate use of antibiotics. There is an ongoing battle between microbes and antibiotic use by humans. This fosters the need for new antibiotics, and for antibiotics to be used more wisely in order to maintain their effectiveness.

Acyl-D-Ala-D-Ala: a precursor of bacterial cell walls. Several bacterial enzymes bind to this molecule and catalyse chemistry at the highlighted carbonyl group.

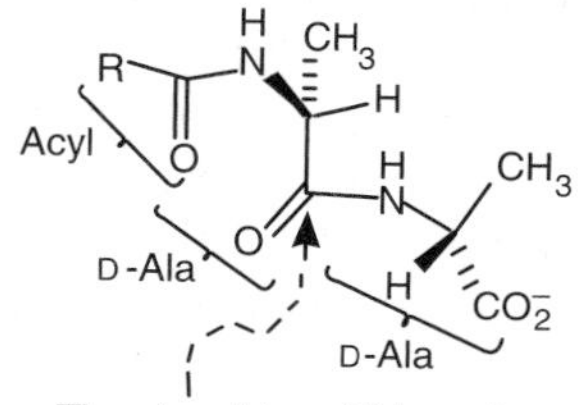

The chemistry of this carbonyl group is important in bacterial cell wall biosynthesis.

Penicillin G, a typical penicillin, resembles acyl-D-Ala-D-Ala and binds to bacterial enzymes which act on this peptide. Once bound, the unusual reactivity of the highlighted carbonyl group can lead to enzyme inhibition.

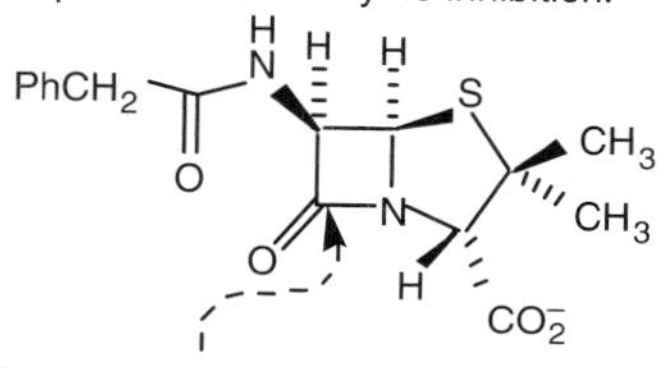

The reactivity of this carbonyl group, incorporated into a strained four-membered ring, is enhanced relative to that of a normal peptide.

An important structural feature of bacterial cell walls is the presence of peptides derived from D-amino acids (a discussion about the L and D stereochemistries of amino acids was presented in Sections 2.4 and 2.5). The use of D- rather than L-amino acids presumably provides a defence against enzymes that rapidly degrade proteins derived from l-amino acids. Penicillins inhibit bacterial enzymes which catalyse reactions of acyl-D-Ala-D-Ala. They are analogues of these D-alanine-derived peptides, but are unusually reactive. Once bound at the target enzymes, they become covalently attached and inhibit further chemistry. The integrity of the bacterial cell wall is undermined and the bacteria die.

Fig. 5.12 Penicillins, important antibiotics, are inhibitors of bacterial enzymes.

5.11 Summary

Triose phosphate isomerase is a well-studied protein whose properties illustrate many of the general features of the way enzymes work. An enzyme is a protein that acts as a catalyst, often dramatically increasing the rate of a specific chemical change by binding a substrate molecule at an active site on its surface. This site provides appropriate functional groups to bring about the reaction. Like many other enzymes, triose phosphate isomerase binds especially tightly to the transition state(s) and high-energy enzyme intermediate(s) involved in the chemical transformation, thereby greatly decreasing the activation energy for reaction. A large number of techniques have been developed to probe the molecular details of enzyme catalysis. Enzyme inhibitors provide information about catalysis and also have important applications, e.g. as pharmaceuticals.

Further reading

Useful texts on enzymology include: A. R. Fersht (1999) *Structure and Mechanism in Protein Science*, W. H. Freeman and Co., New York; and N. C. Price and L. Stevens (1999) *Fundamentals of Enzymology*, 3rd edn, Oxford University Press, Oxford.

A biochemistry text with a strong emphasis on enzyme chemistry is R. H. Abeles, P. A. Frey and W. P. Jencks (1992) *Biochemistry*, Jones and Bartlett Publishers Inc., Boston and London.

Aspects of pharmaceutical chemistry are introduced in: G. L. Patrick (1995) *An Introduction to Medicinal Chemistry*, Oxford University Press, Oxford.

Many of the details of catalysis by triosephosphate isomerase are discussed in a review by J. R. Knowles and W. J. Albery (1977) *Accounts of Chemical Research*, **10**, pp. 105–11; structural information and references to more recent research are reported in R. C. Davenport *et al.* (1991) *Biochemistry*, **30**, pp. 5821–6.

6 Sugars and phosphates: an introduction

6.1 Overview

Chapter 5 gave an introductory account of enzymes, those proteins which catalyse chemical transformations within cells. The particular enzyme highlighted in that chapter, triose phosphate isomerase, catalyses the interconversion of two sugar phosphates. Sugars and their derivatives are important molecules in biochemistry. They are manipulated to provide building blocks for the synthesis of other organic compounds. They are also used as fuel for the generation of cellular energy (and hence act as energy stores), as structural materials, and as components of the molecules of genetic information. An appreciation of the properties of these molecules is therefore essential in understanding the chemistry of cells.

The most important derivatives of sugars in cells are phosphate esters, such as the triose phosphates that were encountered in the last chapter. All the remaining chapters of the book will discuss aspects of the chemistry of such molecules. This chapter will focus on the structural and chemical properties of sugars and their phosphate esters. Succeeding chapters will discuss their biological importance. Chapter 7 outlines the way in which chemical reactions of sugar phosphates are used within cells. Chapter 8 describes derivatives of sugar phosphates that are key constituents of the membranes which surround cells. Finally in Chapter 9, polymeric sugar phosphate derivatives, used to store the genetic information of cells, are described.

6.2 What are sugars?

A simple sugar or 'monosaccharide' can be represented as a straight-chain aldehyde or ketone that bears hydroxyl groups on each of the non-carbonyl carbons. The overall molecular formula of such a compound is $(CH_2O)_n$; hence these molecules are also known as carbohydrates. Sugars and their derivatives are ubiquitous in nature. Common derivatives include esters, and oxidized and reduced forms such as carboxylic acids and polyols, respectively. Figure 6.1 shows the structure of a very simple sugar, glyceraldehyde, and two simple derivatives. As with amino acids, the presence of multiple functional groups in a single molecule allows the possibility of polymerization. Polymers formed by linking individual sugars together are known as polysaccharides and have a wide variety of roles in biology, as will be mentioned in Section 6.4.

Derivatives of sugars include compounds with a wide range of chemical functional groups. As noted in the main text, oxidation of sugars can generate carboxylic acids. Amine groups are present in amino sugars. Hence, the key functional groups of α-amino acids can be found in some sugar derivatives. Polymers and glycoproteins, in which sugars are attached to proteins, are also important in biology, particularly at cellular membranes and surfaces.

The alcohol groups of sugars can form esters with a variety of acids. An ester is simply the product of combining an acid and an alcohol with the loss of water. Esters formed from carboxylic acids and, especially, phosphoric acid predominate in biochemistry.

Glycerol is a component of common lipids and is discussed further in Chapter 8.

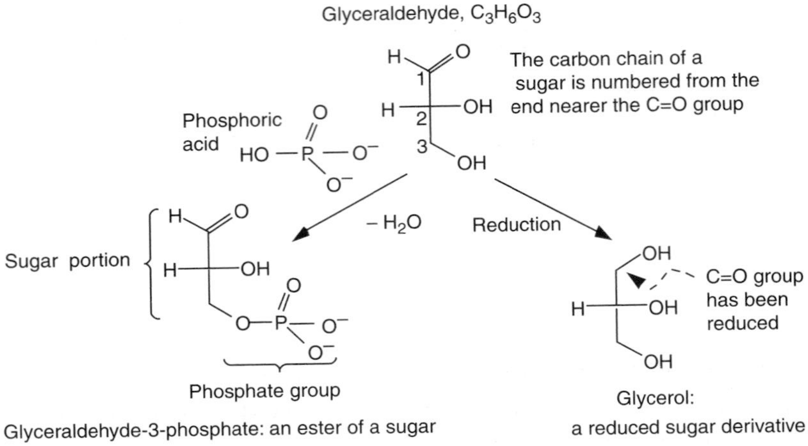

Fig. 6.1   A sugar, glyceraldehyde, and two simple derivatives.

6.3 Understanding the structures of sugars

Most sugars and their derivatives are complex molecules. The plethora of functional groups in a typical sugar often strikes terror into the heart of a student who glances at them, especially if the aim is to memorize the precise structures. It is important for any fully fledged chemical biologist to have an exact knowledge of the structure of biochemically important sugars. In the first instance, however, it is more important to acquire a familiarity with their general nature. As with amino acids and proteins, an appreciation of the biological chemistry of sugars rests on a combination of structural and chemical concepts.

The structural features of sugars will be illustrated by looking firstly at some small, relatively simple examples, e.g. glyceraldehyde and dihydroxyacetone (three-carbon chain, $n = 3$). An analysis of their properties will be extended to discuss ribose ($n = 5$) and glucose ($n = 6$), more complex sugars. Finally, the properties of polymers and phosphate esters of sugars will be considered.

There is a huge variety of molecules which fit the simplest description of a carbohydrate $(CH_2O)_n$. These result from varying three parameters: the length of the carbon chain, n, (up to six carbons is common, although longer sugars are known); the position of the carbonyl group along the chain; and the configuration of the chiral centres associated with each of the internal carbons (other than the carbonyl group).

6.3.1 Trioses: sugars with a three-carbon chain

Glyceraldehyde (Fig. 6.2) comprises a three-carbon chain and hence is described as a *triose*. Since the carbonyl group is in the form of an aldehyde, glyc*eraldehyde* is an *aldose*. The central carbon is a chiral centre; two enantiomers of glyceraldehyde are therefore possible. Sugars in biology are usually found predominantly in only one absolute configuration. As with amino acids, the enantiomers are traditionally labelled L and D; the two enantiomers of glyceraldehyde were arbitrarily assigned as L and D by Fischer and used as the reference compounds for the naming of other metabolites (see Chapter 2).

Dihydroxyacetone, in which the only internal carbon is present as a carbonyl group, is the only sugar with a chain length of 3 or greater which does not contain a chiral centre.

Dihydroxyacetone (Fig. 6.2) is the only other possible triose. It is an isomer of glyceraldehyde in which the carbonyl group resides at the central carbon instead of the terminus. Since it is a ketone, dihydroxyacetone is described as a *ketose*.

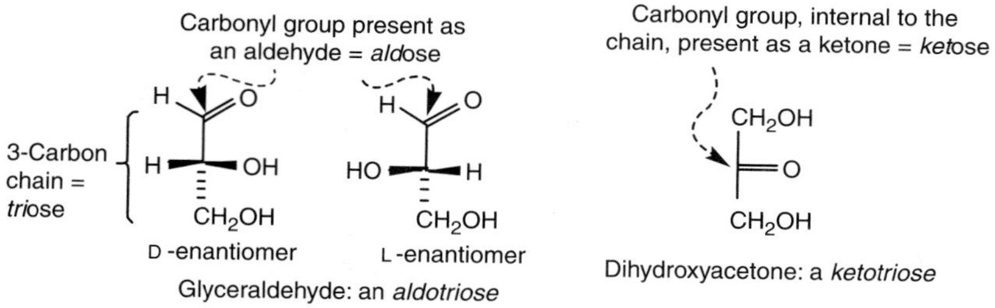

Carbonyl group present as an aldehyde = *ald*ose

Carbonyl group, internal to the chain, present as a ketone = *ket*ose

3-Carbon chain = triose

D -enantiomer   L -enantiomer

Glyceraldehyde: an *aldo*triose

Dihydroxyacetone: a *keto*triose

Fig. 6.2 The structures of three trioses.

6.3.2 Extending the chain: tetroses, pentoses and hexoses

The trioses glyceraldehyde and dihydroxyacetone (Fig. 6.2) illustrate the principles of how isomeric sugars arise by variation in the position of a carbonyl group along the chain, and in the configuration of chiral centres. Another variable parameter is the length of the carbon chain; it is therefore appropriate to consider some possible *tetr*oses which arise by extending the chain by one carbon. The resulting aldoses are similar to glyceraldehyde, but have two chiral centres. Four stereoisomers are possible, corresponding to two pairs of enantiomers; one of these pairs, D- and L-erythrose, is shown below.

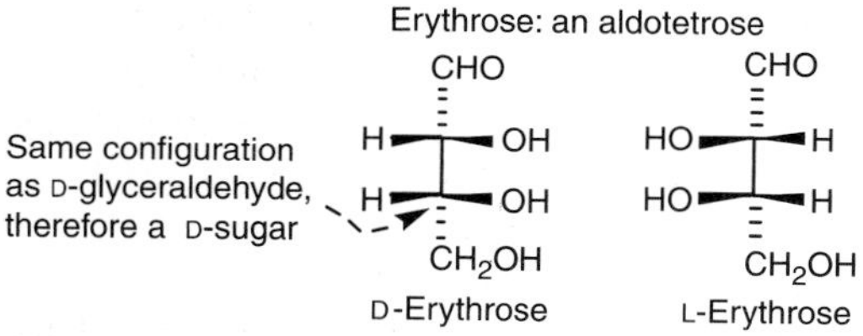

Erythrose: an aldotetrose

Same configuration as D-glyceraldehyde, therefore a D-sugar

D-Erythrose   L-Erythrose

The number of possible isomers increases dramatically as the size of a sugar increases. Biochemistry textbooks generally provide exhaustive information about the names and structures of all the sugars found in nature. At this point it is sufficient to realize that these variations on a simple theme give rise to a wealth of molecular complexity. Most of the important sugars in biochemistry have six or fewer carbons in the chain.

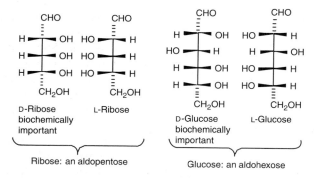

D-Ribose
biochemically
important

L-Ribose

Ribose: an aldopentose

D-Glucose
biochemically
important

L-Glucose

Glucose: an aldohexose

Fig. 6.3 Structures of representative pentoses and hexoses.

Since there are two configurations of a chiral centre, when an additional chiral centre is present in a molecule the number of possible stereoisomers doubles. Sometimes, duplications arise and the total number of isomers is less than expected. In general, if a molecule possesses x chiral centres there are up to 2^x possible stereoisomers.

Given the possibility of varying both chain length and stereochemistry, at internal chiral centres, there is a vast array of sugar molecules potentially available to a biological system. Some forms (e.g. D-glucose) are utilized much more than others.

In addition to L- and D-erythrose, there is another pair of enantiomeric aldotetroses: L- and D-threose. The D-form is shown below.

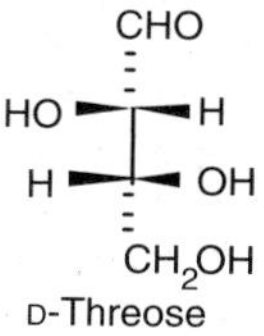

D-Threose

You should be able to draw the L-form by analogy.

As the carbon chain is extended further, additional ketoses also arise, e.g. L- and D-erythrulose are the two enantiomeric ketotetroses.

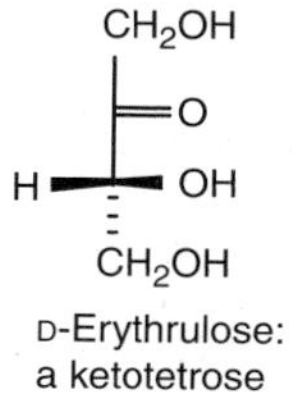

D-Erythrulose:
a ketotetrose

Ribose and glucose are examples of sugars with longer carbon chains, representing a pentose and a hexose, respectively. The D-enantiomers of these sugars are very important in biochemistry and they are discussed further in later chapters. The names ribose and glucose each refer to enantiomeric pairs of stereoisomers with the relative stereochemistries shown in Fig. 6.3. As will be discussed further in Chapter 7, D-glucose and its derivatives are important in the metabolism of all cells. D-Ribose is a component of nucleic acids, molecules associated with the genetic information of cells, and these will be the focus of Chapter 9.

The structures of sugars in aqueous solution

When D-glyceraldehyde is dissolved in water, it does not all remain as a free aldehyde. Carbonyl groups are polarized and the carbonyl carbon is relatively electron deficient. Water is a good nucleophile and can add to the carbonyl group to form a hydrate (Fig. 6.4). In aqueous solution an equilibrium is established between D-glyceraldehyde and its hydrate.

This addition reaction is not restricted to water, and other nucleophiles, such as the hydroxyl group of an alcohol, can react in an analogous fashion. Sugars contain many such hydroxyl groups and so most are prone to undergo internal addition reactions of this type to give cyclic structures. For example, unlike D-glyceraldehyde, only a trace of D-ribose is present as an acyclic molecule under equilibrium conditions. Figure 6.5 shows how the linear structure of ribose can be redrawn to emphasize the ease with which it cyclizes to form a five-membered ring.

The formation of a hydrate, as with related reactions described later, is freely reversible and is catalysed by both acids and bases. Figure 6.4 illustrates the mechanism of this reaction when catalysed by acid.

Because of the preferred bond angles of carbon and oxygen, small rings, i.e. those with fewer than five atoms in the ring, are strained. By contrast, five- and six-membered rings are the least strained ring structures. Sugars of chain length 4 or greater, which are capable of forming such rings, tend to cyclize.

In the cyclization reaction, the carbonyl group is converted into a 'hemi-acetal' group. The carbon of this group is sometimes called an 'anomeric centre'. The anomeric carbon is actually a chiral centre and so two stereoisomers are possible; these are termed anomers. Since hemi-acetal formation is freely reversible in the cell, the two anomers are readily interconverted. The isomer in which the new hydroxyl group lies below the plane of the ring, as drawn, is called the α-anomer; the other isomer is known as the β-anomer.

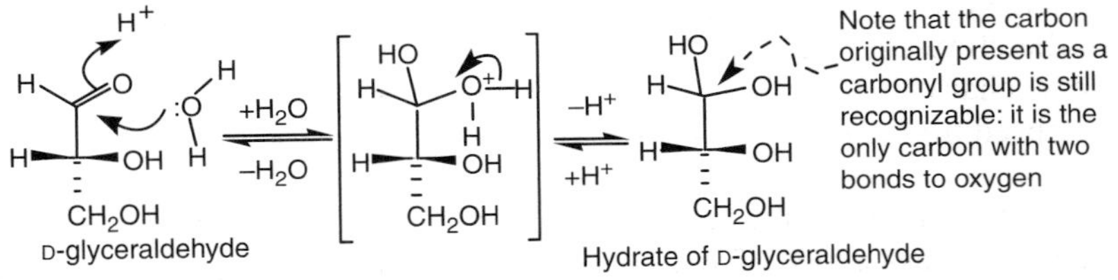

Note that the carbon originally present as a carbonyl group is still recognizable: it is the only carbon with two bonds to oxygen

Fig. 6.4 Hydration of D-glyceraldehyde.

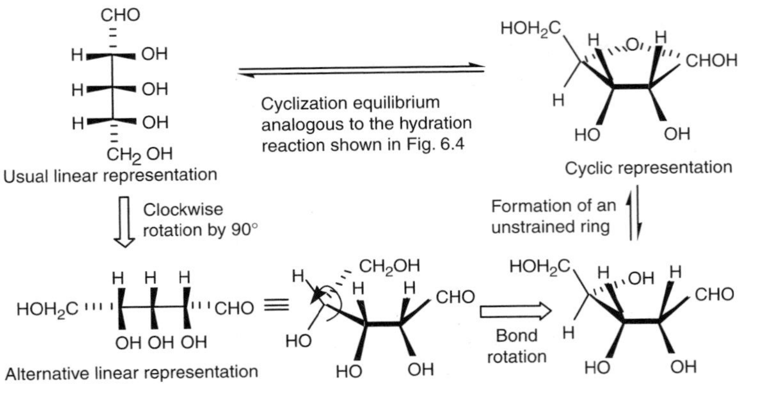

Fig. 6.5 Visualizing the relationship between linear and cyclic forms of ribose.

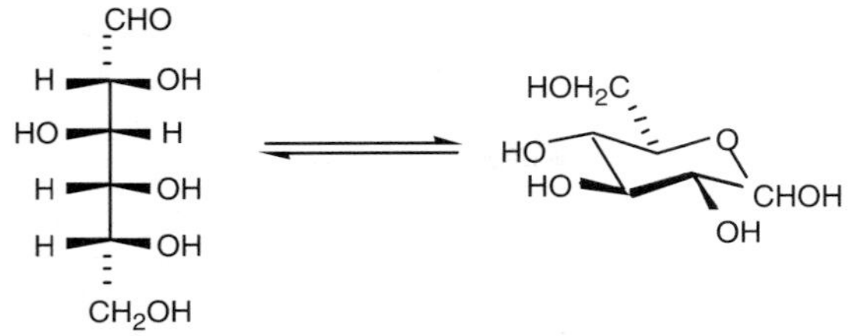

Fig. 6.6 Cyclization of D-glucose.

Glucose can adopt a 'chair' conformation, as shown in Fig. 6.6, in which all the substituents (other than at the anomeric centre) are in an 'equatorial' position. It may be that the thermodynamic stability of this arrangement, where steric clashes are minimized, is one of the reasons why glucose is the most common hexose in nature.

Likewise, glucose exists primarily in cyclic forms. In this case, a six-membered ring is the preferred structure. This is illustrated in Fig. 6.6.

6.4 Polymeric derivatives of sugars

As with amino acids, sugars are elaborated into polymers by dehydration chemistry. Since these polymers are of great biochemical significance, their properties will be outlined briefly here. After initial analysis of a dimer, maltose, discussion will be extended to the macromolecules starch and cellulose. Maltose is a common disaccharide produced, for example, in the malting process during beer production. It is derived from glucose (Fig. 6.7). It corresponds to reaction of one of the alcohol groups of glucose (that on C-4) with the carbonyl carbon (C-1) of a second glucose. The formation of maltose from glucose is a dehydration reaction, so the reverse reaction, liberating glucose from maltose, is a hydrolytic process. In an aqueous environment, the hydrolysis of oligosaccharides to the component monomeric sugars is thermodynamically favourable (maltose can therefore act as a store of glucose). Nonetheless, oligosaccharides and polysaccharides are kinetically stable in the absence of specific enzymes that speed up the hydrolysis reaction, and they serve important functions in cells.

As with amino acids, the formation of polymers from sugars can generate a vast array of macromolecules. Different monomers can be

Short biological polymers are often named oligomers. Oligomer is a term derived from the Greek 'oligo', meaning few. As an example, relatively short polymers derived from sugars are called oligosaccharides, whereas longer polymers are called polysaccharides. This distinction parallels the difference between the names peptide and protein.

The formation of maltose from glucose is a nucleophilic substitution reaction. It is formally the displacement of hydroxide from the C-1 position of one glucose by a nucleophilic hydroxyl group at the C-4 position of the other glucose. This is an acid-catalysed reaction taking place via an S_N1 mechanism. Protonation of the hydroxyl of C-1 (to produce a good leaving group) and loss of water generates a relatively stabilized carbocation. Nucleophilic attack on this ion, by the C-4-hydroxyl group of a second glucose molecule, generates the 'acetal' linkage between the two glucose mono-mers. It is a useful exercise to draw out this mechanism in full.

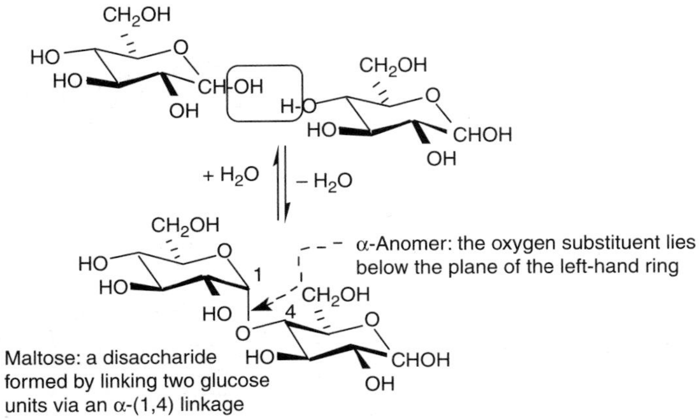

In the β-anomer the oxygen substituent is above the plane of the ring. The structure of cellulose in Fig. 6.8 provides an example.

Maltose: a disaccharide formed by linking two glucose units via an α-(1,4) linkage

Fig. 6.7 Maltose: an example of a disaccharide.

Sugar derivatives, especially glycoproteins (macromolecules made from a combination of proteins and polysaccharides), are important in molecular recognition processes occurring at the surface of cells (see Chapter 8).

Acetals are less reactive than hemi-acetals (both are sensitive to acids but only hemi-acetals are sensitive to base). Acetal linkages are stable in the cell in the absence of enzymes able to catalyse their hydrolysis. Hence, in contrast to hemi-acetals, the configuration at the anomeric centre of an acetal is stable in the cell. Starch and cellulose, which differ in the configuration at this centre, are, therefore, distinct molecules.

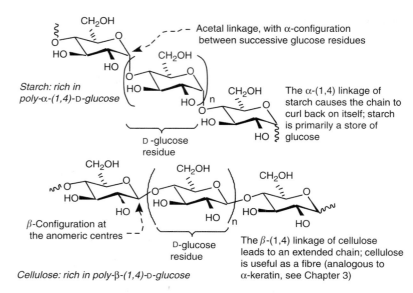

Fig. 6.8 Starch and cellulose: two polysaccharides.

incorporated into the growing chain; the stereochemistry of the linking anomeric centre can be in either of two configurations; and the presence of multiple hydroxyl groups within sugars allows a variety of linkages to be made and facilitates branching (in cases where more than one of the hydroxyl groups of a sugar monomer is used to form an acetal linkage). With this structural diversity available, it is not surprising that polysaccharides are involved in a wide variety of biochemical roles.

Despite their possible complexity, many polysaccharides involve simple repeating patterns. So, for example, two of the most common polysaccharides, starch and cellulose, are primarily polymers of glucose which are linked by acetal groups between the C-1 and C-4 substituents of glucose monomers (Fig. 6.8).

Cellulose has an extended rod-like structure and it is used as a structural material, notably as a fibrous constituent of plant tissue. Amylose, a polysaccharide component of starch, is used as a store of chemical energy, e.g. plant tubers, such as potatoes, are rich in this polymer. This is a helical polymer (see Fig. 6.9) which is broken down into glucose which, in turn, is oxidized to liberate chemical energy when the plant grows (see Chapter 7).

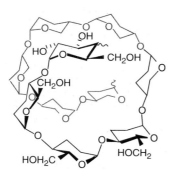

Fig. 6.9 Representation of the structure of starch amylose (some groups are omitted for clarity).

In the structure of amylose the chain twists in a constant way at each residue. This results in a helical structure in an analogous fashion to those found in protein secondary structures, such as α-helices.

6.5 Sugar phosphates

Another important class of sugar derivatives, of great biochemical significance, is that of the phosphate esters. Their biochemical utilization can be understood with reference to a few fundamental chemical properties, which are listed as headings below. The biological significance of these properties will emerge in more detail in subsequent chapters.

There are three types of phosphate ester

Phosphoric acid has three acidic protons. Hence phosphoric acids can form three different types of ester: monoesters, diesters and triesters (the latter type is not important in biological systems). The ability to form diesters allows sugar phosphates to form polymeric chains.

Phosphoric acid · Phosphate monoester · Phosphate diester · Phosphate triester (not biochemically important)

Simple phosphates are anionic in cells

Phosphoric acid and simple derivatives of it that retain at least a single hydroxyl group (e.g. monoesters and diesters) are all strong acids. They are found as anions under physiological conditions. The ionic nature of phosphates has several ramifications. Incorporation of a phosphate group into an organic compound usually increases its solubility in water. Since every cell is separated from the outside world by a non-polar membrane, which cannot easily be crossed by ions, simple phosphates are usually retained within cells.

Further deprotonation can occur if an R group is hydrogen

Anionic forms predominate when pH > 2

Phosphoric acid is a tribasic acid:

$$H_3PO_4$$
$$+H^+ \updownarrow -H^+ \quad pK_a \sim 2$$
$$H_2PO_4^-$$
$$+H^+ \updownarrow -H^+ \quad pK_a \sim 7$$
$$HPO_4^{2-}$$
$$+H^+ \updownarrow -H^+ \quad pK_a \sim 12$$
$$PO_4^{3-}$$

Since $pK_a = pH$ at the mid-point of equilibrium, phosphates are only fully protonated at pH values significantly lower than 2.

Phosphates, as anions, can interact strongly with molecules through electrostatic forces. These interactions provide useful sources of binding energy and molecular selectivity.

Phosphate derivatives hydrolyse slowly

The formation of esters from acids and alcohols is a reversible process. In fact the reverse reaction, hydrolysis, is thermodynamically favoured in water. However, in the absence of catalysis, most phosphate esters and related compounds hydrolyse only slowly, and once made are essentially stable under physiological conditions. They can, therefore, be used by cells and broken down when appropriate.

There is electrostatic repulsion between the electrons of the attacking nucleophile and those of the anionic oxygen

The electrostatic repulsion between the oxygen atom of water and the negative charge of a phosphate group raises the energy of the transition state required for nucleophilic substitution; this provides a rationalization for the relative slowness of the hydrolysis of phosphate derivatives. Divalent cations, e.g. Mg^{2+}, are frequently used to speed up biochemical phosphate chemistry by diminishing electrostatic repulsion.

Phosphates are good leaving groups

Many of the important reactions of biological chemistry are nucleophilic substitution and elimination reactions. For these reactions to be facile, a 'good leaving group' is required. Phosphates are widely employed in

A good leaving group is typically the conjugate base of a strong acid (e.g. hydrochloric acid is a strong acid and the chloride ion is a good leaving group; see the discussion in Chapter 2).

A combination of acid and base catalysis can bring about the migration of the carbonyl group along the carbon chain (discussed in detail in Chapter 5), the interconversion of configurations of chiral centres, and changes in the length of the carbon chain of a sugar (chains can be cleaved to form smaller sugars and, conversely, smaller chains can be assembled into larger sugars). Although some of the details of this chemistry are discussed in other chapters, a thorough analysis is beyond the scope of this book. An excellent account is given in *Organic Chemistry* by Kemp and Vellacio (see Further reading).

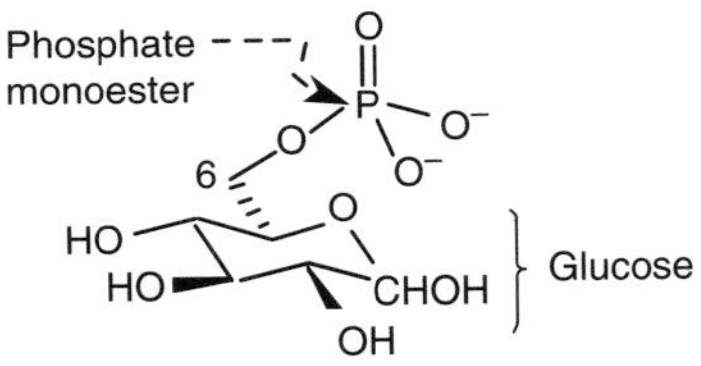

D-Glucose-6-phosphate: an important derivative of glucose (see Chapter 7).

Fig. 6.10 A representative sugar phosphate.

biochemistry when good leaving groups are required (e.g. in protein synthesis, see Section 7.2).

6.6 The importance of sugars and derivatives

Why does biology make so much use of sugars? Sugars are available in the biosphere through the reduction of carbon dioxide, the most plentiful inorganic carbon source. Importantly, sugars can be readily inter-converted into other sugars when appropriate catalysts are available. Enzymes are well equipped for this role, as was described for one such example, triose phosphate isomerase, in Chapter 5. In short, access to appropriate enzymes makes it possible for cells to convert virtually any sugar, or simple derivative, into any other sugar or related derivative. Sugars are molecules with a rich variety of chemistry which are capable of acting as a universal currency in the biochemical economy.

Most monomeric sugars are handled within cells as phosphate ester derivatives (Fig. 6.10). The anionic phosphate group ensures substantial water solubility, prevents loss of the sugar phosphate through the cell membrane, provides a 'handle' for recognition by proteins, and is a reactive chemical functional group which can be exploited when necessary.

6.7 Summary

At one level, sugars are simple molecules: linear chains of carbon atoms including a carbonyl group, having hydroxyl groups on the other carbons. This simple description quickly becomes complex as the chain increases in length, due to variations in the position of the carbonyl group and the stereochemistry of chiral centres. The complexity is increased by the wide range of simple derivatives of sugars that is possible. The combination of functional groups in sugars and their derivatives allows a rich range of organic transformations to be carried out under mild conditions. The challenge of sugar chemistry is to perform the desired transformations in a selective fashion, when so many alternatives exist. Simple acid- and base-catalysis can interconvert sugars, but not in a controlled way. Cells produce enzymes which increase the rate of specific reactions relative to others; they can therefore readily convert a particular sugar to a single product at will by employing an appropriate enzyme. Sugars are a universal biochemical currency. Selective reactions of sugars and their derivatives can provide cells with a variety of organic building blocks and can be used to generate chemical energy. These principles will be illustrated with reference to some aspects of glucose biochemistry in the next chapter.

Further reading

D. S. Kemp and F. Vellacio (1980) *Organic Chemistry*, Worth Publishers, Inc., New York, contains some excellent discussions of the chemistry of sugars and related compounds, especially in chapters 25 and 27.

Another general textbook with a good coverage of sugars is M. Jones Jr (1997) *Organic Chemistry*, W. W. Norton & Co., New York, especially chapter 24.

F. H. Westheimer (1987) Why Nature Chose Phosphates, *Science*, **235**, pp. 1173–8, analyses the exploitation of phosphates in biochemistry.

7 Metabolism and the biochemistry of glucose

7.1 Overview

Cells continually degrade organic compounds and synthesize new ones. The breakdown of complex compounds provides simple organic building blocks for the synthesis of other biological compounds as they are required. Chemical transformations of simple organic compounds are also used by cells to generate energy in a usable form. These chemical inter-conversions of 'metabolites' are the basis of metabolism. Of all the metabolites involved in these transformations, sugars and their simple derivatives play a central role.

The biochemical breakdown of organic compounds into simpler species is often termed catabolism, whereas the synthesis of complex molecules from simpler precursors is known as anabolism.

This chapter describes the nature of chemical energy within cells and how this energy is harnessed from enzyme-catalysed reactions. Cellular energy is related to the ability to facilitate dehydration chemistry in aqueous solution and the primary source of dehydrating power in cells, a phosphate derivative (ATP), is introduced.

Having established the nature of biochemical energy, the biochemistry of glucose is used as an example of the manipulation of chemicals in cells: degradation to common building blocks; use of these building blocks to prepare other chemicals; and the generation of biochemical energy. Some features of coenzyme chemistry are also introduced in this discussion.

7.2 The chemical energy of cells

Cells use many forms of energy, including mechanical energy (e.g. in muscle contraction) and electrical energy (e.g. in nerve signals), as well as chemical energy. However, all the energy-requiring processes of cells are related to chemical energy and this will provide the focus of our discussion. The nature of chemical energy within cells relates to the ability to turn processes that would normally be unfavourable into favourable ones; it is easiest to understand this concept by the use of examples.

Precisely constructed polymers are characteristic of life. The polymers common to all living systems correspond to the joining of monomer units with concomitant loss of water. They are known as *condensation polymers*, and include proteins derived from amino acids (shown schematically in Fig. 2.2, see Section 2.1), polysaccharides derived from sugars (introduced in Section 6.4) and nucleic acids, the molecules associated with genetic information, which will be described in Chapter 9.

To a considerable extent, we can identify the capacity to carry out controlled dehydration reactions as a key chemical feature of cells. Since cells are approximately 70 per cent water, dehydration chemistry is

Chemical energy is used to drive
all energy-dependent processes
in biology. For example,
mechanical energy, as
exemplified by moving muscle,
actually corresponds to moving
protein fibres relative to one
another, and is brought about by
cycles of dehydration and
hydrolysis chemistry.

generally unfavourable. In dilute aqueous solution the *hydrolysis* of
condensation polymers is the more thermodynamically favourable process. When we digest meat (mainly protein) and potatoes (mainly polysaccharides), we catalyse the favourable hydrolysis of polymers into the
corresponding monomers in an aqueous environment. It is the construction of polymers from these monomers that requires energy.

In order for a dehydration reaction to occur in a cell, the overall process
must be made *thermodynamically* favourable; and the rate of reaction
must be appropriate at the temperature found within cells. Enzymes can
be used to address the latter (*kinetic*) issue but not the former, as they alter
the rate of attainment, but not the position, of equilibrium. A molecule
that can be used by cells to make a dehydration process thermodynamically favourable in an aqueous environment has, however, the
potential to act as a source of chemical energy.

7.2.1 An important biochemical dehydrating agent: ATP

Adenosine, and related
phosphate esters (nucleotides),
act as 'handles' for a number of
enzyme cofactors. The structures
of adenosine and related
molecules are analysed in more
detail in Chapter 9.

The molecule most commonly identified as a source of chemical energy
within cells is an acid anhydride, adenosine triphosphate, commonly
abbreviated to ATP. ATP is a complex molecule but only a certain portion
of the molecule is associated with its role as a store of chemical energy. It is
the triphosphate portion, a phosphate anhydride group (Fig. 7.1), which is
a dehydrating agent; the remainder of the molecule, adenosine, can be
thought of as a 'handle' to carry this functional group, facilitating effective binding by enzymes.

The three phosphate centres are
denoted by α, β and γ, in order of
increasing separation from the
adenosine unit–see Fig. 7.2. In
metabolic reactions, nucleophilic
substitution occurs most
commonly at the γ- and
α-phosphates.

ATP can react with nucleophiles at any of the phosphate centres.
Figure 7.2 shows a nucleophilic substitution at the γ-phosphate. Such
reactions, e.g. hydrolysis, result in the displacement of a good leaving
group—a phosphate derivative, in this case *adenosine diphosphate*
(ADP). A key feature of phosphoric anhydrides is that nucleophilic substitution reactions are energetically *favourable*; however they generally
occur *slowly* (see Section 6.5) in the absence of a catalyst.

The fact that phosphoric anhydrides are thermodynamically unstable
but kinetically stable allows them to be maintained in relatively high
concentrations in cells. Consequently their reactions with a variety of
nucleophiles are controlled by specific enzymes.

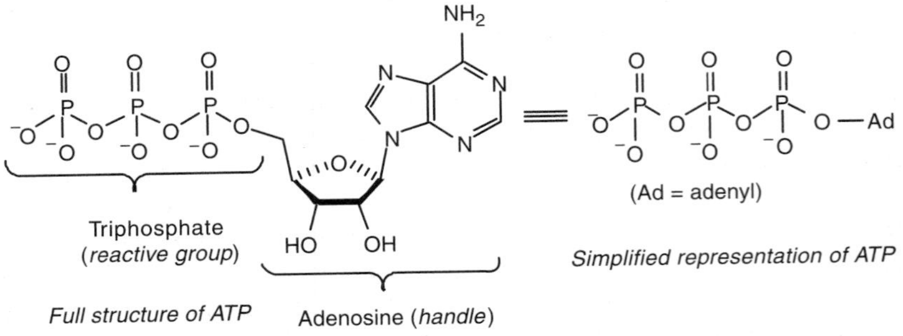

Fig. 7.1 ATP: a phosphoric acid anhydride.

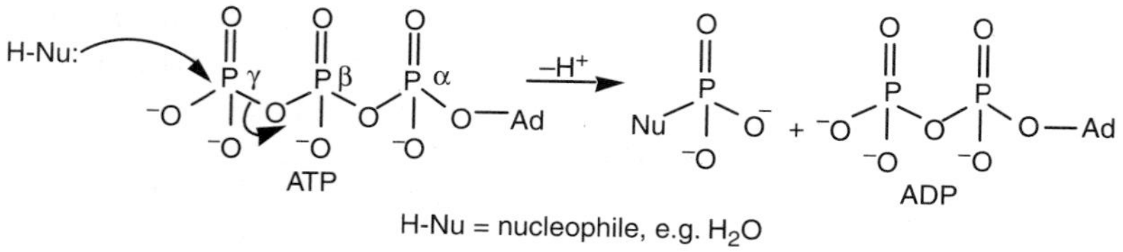

H-Nu = nucleophile, e.g. H_2O

Fig. 7.2 Overview of a nucleophilic substitution at ATP.

As mentioned in Section 6.5, one factor that slows the rate of nucleophilic substitution at phosphate derivatives is the electrostatic repulsion between the attacking lone pair of electrons on the nucleophile and the anionic oxygens of the phosphate group.

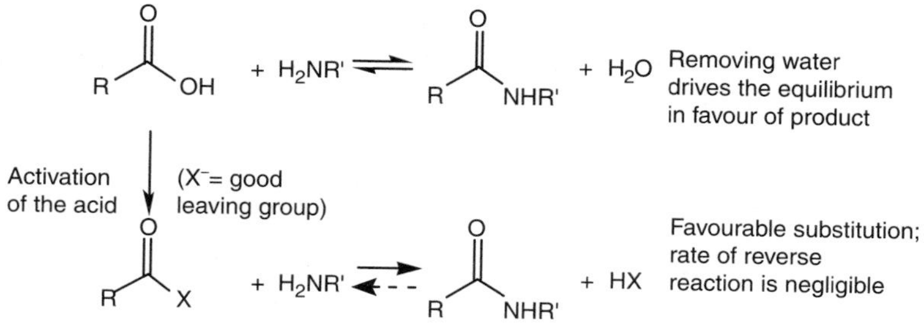

Removing water drives the equilibrium in favour of product

Activation of the acid | (X^- = good leaving group)

Favourable substitution; rate of reverse reaction is negligible

Fig. 7.3 Amide bond formation—two alternatives.

The fact that proteins are condensation polymers was introduced in Section 2.1.

7.2.2 Protein biosynthesis: an example of controlled dehydration in cells

The production of a wide array of proteins is essential to cellular function. The synthesis of proteins involves controlled addition of amino acids to a growing polypeptide chain, corresponding to the elimination of one molecule of water for each amide bond formed (Fig. 2.2). Faced with the challenge of making amide bonds from amines and acids, chemists manipulate the position of the equilibrium by one of two strategies: changing the concentration of one of the species involved, notably removing water as it is formed; or making a derivative of the acid, e.g. an anhydride or an ester, such that subsequent formation of the amide is favourable (illustrated schematically in Fig. 7.3). The first option is not possible in the aqueous environment of the cell. The second approach, which circumvents the production of free water, *is* viable in cells and is the one adopted by nature: a carboxylic acid is converted into a derivative which has a better leaving group.

The first stage in the biosynthesis of proteins is the reaction of an amino acid with ATP. Nucleophilic attack by the carboxylate group at the α-phosphate centre generates an acyl phosphate (Fig. 7.4). This acid derivative, a mixed acid anhydride, is activated with respect to nucleophilic substitution: the carbonyl centre now bears a good leaving group. The acyl phosphate produced is rapidly converted to an ester, an intermediate which is relatively stable (Fig. 7.4). The ester is formed with an alcohol of complex structure, a polymer known as a 'transfer RNA', or tRNA. This class of molecule will be discussed in more detail in Chapter 9.

The esters formed between the tRNA and the amino acid are then transformed into proteins by a favourable series of nucleophilic substitution reactions (Fig. 7.5).

The biosynthesis of proteins takes place on ribosomes, complex assemblies of ribonucleic acid and a variety of proteins. The basis of ribosomal synthesis of proteins is discussed in Section 9.11. The discussion in this section focuses on the background chemical issues associated with peptide bond formation.

In the two-stage process, involving the activation of the acid, water is not produced. Instead, it forms part of the leaving group that is displaced by the amine nucleophile.

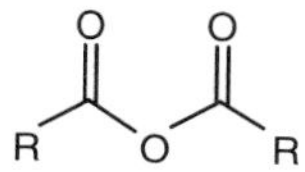

Simple carboxylic acid anhydrides (see above) react efficiently with amines to form amides, but they hydrolyse readily in water and are not used as dehydrating agents in cells.

Amino acid activation involves attack at the α-phosphate, rather than the γ-phosphate, of ATP. The overall reaction is more favourable because of the ready subsequent hydrolysis of the other reaction product: pyrophosphate. Formation of acyl phosphates from ATP and carboxylates is energetically unfavourable; under normal conditions, the back reaction would predominate. The hydrolysis of 1 mol of pyrophosphate to 2 mol of phosphate is catalysed efficiently by cells. This reaction is strongly favourable; pyrophosphate is destroyed as it is formed and is, therefore, not available for the back reaction.

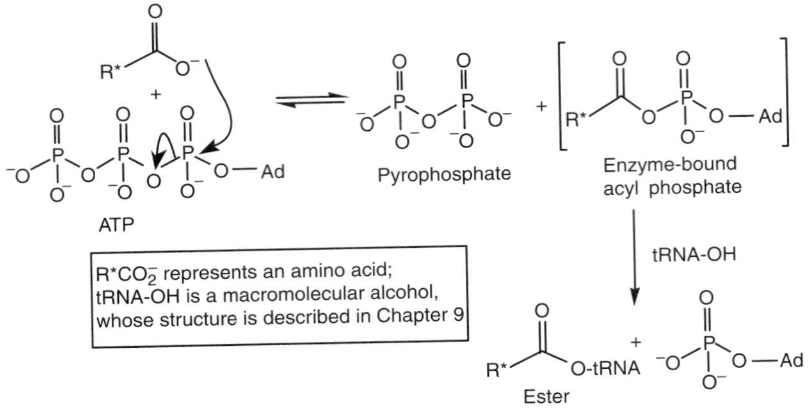

$R^*CO_2^-$ represents an amino acid; tRNA-OH is a macromolecular alcohol, whose structure is described in Chapter 9

Fig. 7.4 The two-step formation of an activated amino acid.

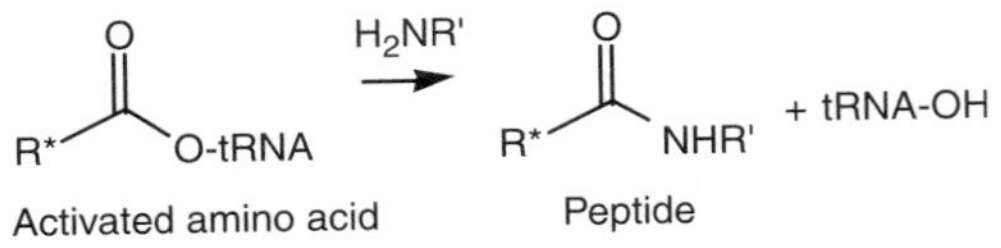

The peptide bond formation process shown in Fig. 7.5 corresponds to the generation of a growing polypeptide chain on the ribosome. The overall process is represented schematically in Fig. 9.21. Suggest a mechanism for this nucleophilic substitution reaction.

Fig 7.5 Peptide bond formation by nucleophilic substitution of tRNA esters.

Overall, the formation of each amide bond is associated with the hydrolysis of a molecule of ATP. This process removes the water that would otherwise have to be displaced. Hence, ATP fulfils the role of a water-compatible dehydrating agent. When cells undertake processes to generate energy, this typically corresponds to the production of ATP. This process, and other aspects of metabolism, will now be illustrated by considering some of the biochemistry of glucose.

7.3 Overview of the catabolism of glucose

All the chiral sugars involved in the catabolism of glucose have the D-configuration (see Chapter 6). In this chapter, for simplicity, the stereochemistry of the sugars is not explicitly labelled.

Chemical reactions of sugars, and their derivatives, are a central feature of metabolism. As an example, glucose is converted to other sugars and derivatives in cells as a means of generating chemical energy. The oxidation of glucose by oxygen, to generate carbon dioxide and water, is strongly exothermic (Fig. 7.6). Many cells in oxidizing environments derive energy by 'burning' sugars; the energy is harvested in a controlled fashion, yielding reactive chemicals rather than heat.

Controlled oxidation is accomplished by carrying out the breakdown of glucose via a series of discrete chemical reactions, each of which is mediated by a particular enzyme. These reactions lead to the progressive shortening of the carbon chain. Reactive chemicals, ATP in particular, are generated by this oxidative chemistry and can be used when required to bring about chemical reactions in the cell which would otherwise be unfavourable. The small molecules that result from the breakdown of the sugar can also be used as building blocks for the construction of other chemical compounds.

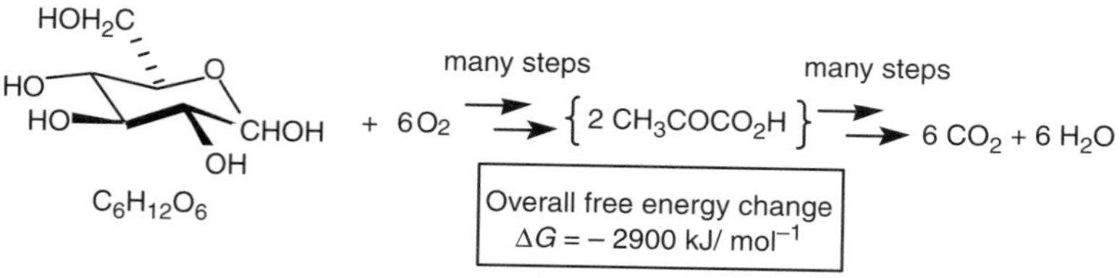

Fig. 7.6 Oxidative breakdown of glucose.

7.4 The oxidative metabolism of glucose: glycolysis

The breakdown of glucose to a three-carbon acid, pyruvate, is known as glycolysis. It is useful to consider the strategy underlying the overall process before turning to the specific details. Firstly, the cell takes up glucose from its surroundings and retains it by phosphorylating it; secondly, the six-carbon chain of the sugar is cleaved to generate two equivalent three-carbon units; finally the three-carbon fragments are manipulated to generate useful building blocks and chemical energy in the form of ATP. These three stages are described in the following three subsections.

7.4.1 Glucose uptake by cells

Glucose is readily taken up by many cells. In the absence of any other process, glucose would flow into a cell until the concentrations inside and outside the cell are the same. It is obviously of benefit to cells to draw in as much glucose as possible (ideally all of the available glucose) and retain it. Cells achieve both aims by making a phosphate ester of glucose, glucose-6-phosphate, inside the cell.

Glucose is phosphorylated by ATP in a reaction catalysed by the enzyme hexokinase. This process, the formation of an ester with displacement of a phosphate leaving group (ADP), is favourable and essentially all the glucose within the cell is phosphorylated (Fig. 7.7). This means that the concentration of free glucose within the cell is kept close to zero and a concentration gradient favouring entry of more glucose into the cell is maintained.

Not only does any available glucose enter the cell, but once phosphorylated, it is in an ionic form, unable to flow freely through the cell

The oxidative metabolism of glucose to carbon dioxide (by glycolysis and the tricarboxylic acid cycle, as described in this chapter and in Chapter 8) produces up to 38 mol of ATP (from ADP and phosphate) per mol of glucose. The free energy of hydrolysis of ATP is approximately -36 kJ mol^{-1}. Hence the chemical energy harvested, in the form of the dehydrating agent ATP, is approximately $-1350 \text{ kJ mol}^{-1}$ of glucose. This is more than 45 per cent of the maximum theoretical amount. Judged against the standards of most man-made devices, humans utilize energy very efficiently; the average adult dietary 'calorific' intake is about 10 000 kJ (2400 kcal) per day. This corresponds to an average energy consumption of approximately 120 W, about the same as a light bulb.

Enzymes which catalyse the attachment of phosphate groups to organic molecules are known as 'kinases'. Hexokinase is so named because it undertakes this chemistry on *hexoses*.

The structures of ATP and ADP are shown in Figs 7.2 and 7.3.

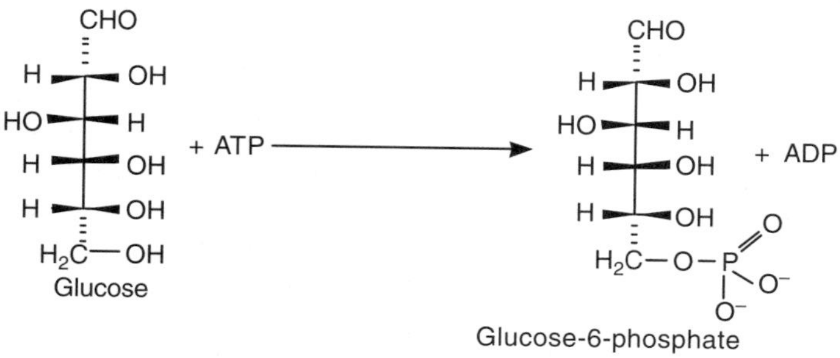

In this chapter sugars are represented as straight-chain structures. This is done to make it easier to follow the chemical transformations. It is important to remember that these are often not the predominant forms of these sugars in solution (see Chapter 6).

The cell membrane has a non-polar core, which provides a barrier to the movement of ionic species (see Chapter 8).

Fig. 7.7 Phosphorylation of glucose.

Many processes by which small molecules and ions pass through cell membranes are 'coupled' to ATP chemistry. Either ATP is consumed as the cell generates a concentration gradient, or ATP is generated as ions flow across a membrane to redress a concentration imbalance.(see Sections 8.4 and 8.5).

Based on the material in Chapter 5, suggest a mechanism for the interconversion of glucose-6-phosphate and fructose-6-phosphate.

Phosphorylation of fructose-6-phosphate to fructose-1,6-bisphosphate is the point at which the cell expends energy (ATP) to commit these molecules to glycolysis. The activity of the enzyme that catalyses this step, phosphofructokinase, is regulated in many cells so that it is only highly active when the cell requires more glycolysis to take place. In the presence of high levels of PEP, a product of glycolysis (see Fig. 7.13), the activity of this enzyme drops. This control is mediated by allosteric effects (see margin notes pp 31 and 37). PEP binds to the surface of phosphofructokinase, stabilizing a structure with decreased catalytic activity.

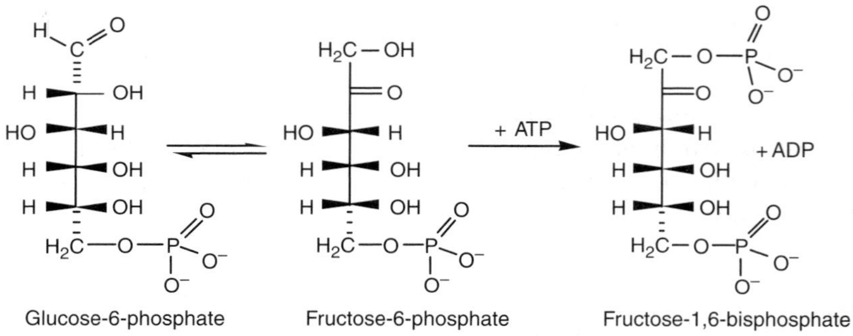

Isomerization of a glucose skeleton into a fructose unit places the carbonyl group in a position where carbon–carbon bond cleavage generates two three-carbon sugars. Carbon–carbon bonds can be made and broken by simple acid–base chemistry when they are β- to a carbonyl group. This chemistry is known as aldol chemistry.

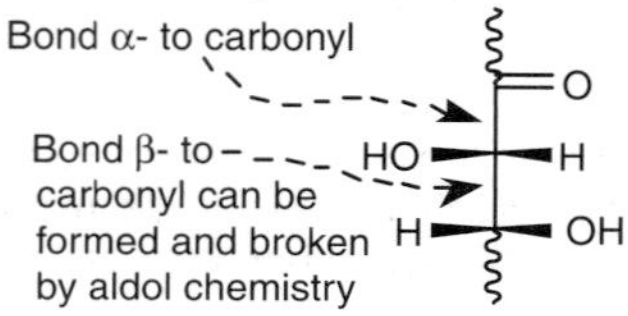

Bond α- to carbonyl

Bond β- to carbonyl can be formed and broken by aldol chemistry

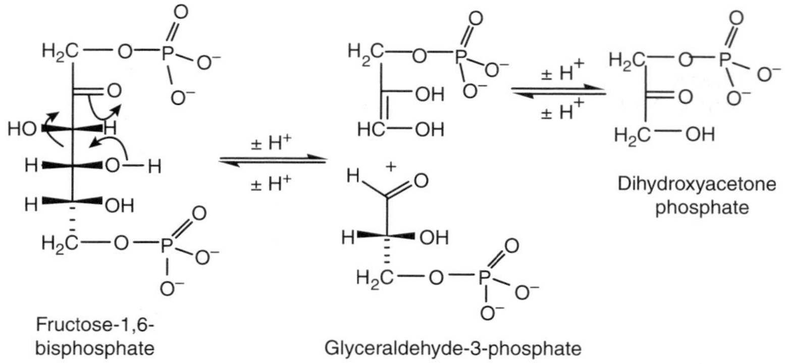

membrane, and hence it is trapped. This illustrates how the chemical energy of ATP can be used to manipulate concentration gradients in a fashion useful to cells.

7.4.2 Fragmentation of the carbon skeleton

After glucose has entered a cell and been phosphorylated to glucose-6-phosphate, it is broken down into smaller organic molecules. In overall terms, this six-carbon sugar is converted into two molecules of a three-carbon sugar, glyceraldehyde-3-phosphate, which is subsequently oxidized; the net result of the oxidation process is the production of ATP. All the intermediates in this pathway are anionic and so are retained by cells.

In the first stage, the carbonyl group of glucose-6-phosphate is moved along the chain. This isomerization generates the compound fructose-6-phosphate. A phosphate group is then added to the other end of the chain, producing fructose-1,6-bisphosphate (Fig. 7.8). With phosphate groups on each end of the chain, fragmentation can generate two smaller molecules, each of which is anionic.

Fructose-1,6-bisphosphate is in fact split into two phosphorylated trioses: glyceraldehyde-3-phosphate and dihydroxyacetone phosphate (Fig. 7.9).

Fig. 7.8 Conversion of glucose-6-phosphate to fructose-1,6-bisphosphate.

Fructose-1,6-bisphosphate

Glyceraldehyde-3-phosphate

Dihydroxyacetone phosphate

Overall reaction: fructose-1,6-bisphosphate $\rightleftharpoons$ dihydroxyacetone phosphate + glyceraldehyde-3-phosphate

Fig. 7.9 Fragmentation of the six-carbon chain into two three-carbon units.

Although two isomeric triose phosphates are produced, only one of these, glyceraldehyde-3-phosphate, is subsequently oxidized. For complete oxidation of glucose, therefore, the dihydroxyacetone phosphate must be converted into glyceraldehyde-3-phosphate. This interconversion is catalysed by TIM and was discussed in detail in Chapter 5. This combination of chemical reactions has succeeded in converting the six-carbon sugar into two equivalent smaller units in readiness for further manipulation.

7.4.3 Energy production

The reactions described so far have not involved any oxidative chemistry and have actually used up cellular ATP, rather than generated it! It is the oxidation of glyceraldehyde-3-phosphate that leads to the production of the first batch of ATP molecules from glucose metabolism. This process is catalysed by glyceraldehyde-3-phosphate dehydrogenase.

The side chain functional groups of proteins provide a variety of acid–base and hydrogen bonding potential, but only limited redox chemistry (notably cysteine/cystine interconversion, see Section 2.3). The latter is not sufficient for all the redox needs of cells. For more complex redox reactions, an organic molecule, a coenzyme, can be conscripted to carry out the oxidation chemistry. In the case of glyceraldehyde-3-phosphate dehydrogenase, this coenzyme, nicotinamide adenine dinucleotide or NAD^+, is a complex pyridine derivative (Fig. 7.10). It is the chemistry of this pyridine group that is utilized by the enzyme.

The oxidation of glyceraldehyde-3-phosphate does not lead to the corresponding acid directly; instead, the reaction product is an activated acid derivative, an acyl phosphate (Fig. 7.11).

The aldehyde group of the aldose glyceraldehyde-3-phosphate can be readily oxidized. By contrast, dihydroxyacetone phosphate is a ketose and ketone groups are much harder to oxidize.

If the C–C bond cleavage chemistry were carried out on glucose-6-phosphate, it would generate a four-carbon sugar phosphate, erythrose-4-phosphate, and a two-carbon sugar, glycolaldehyde.

Note that a wide variety of sugars can be converted into the triose phosphates, which can then be metabolized in a common fashion. This illustrates the biochemical versatility of sugars and the benefit of generating common metabolites.

Cells exploit a range of chemistry not intrinsically present in simple proteins by utilizing other chemical species to provide the extra chemical diversity. An example, discussed in Chapter 4, was the use of a cofactor by myoglobin and haemoglobin. These proteins both utilize Fe^{2+}, encased in a haem group, to bind oxygen. Key building blocks for the construction of coenzymes are often vitamins: essential trace nutrients in the diet. Nicotinic acid (niacin), incorporated into nicotinamides, falls into this category:

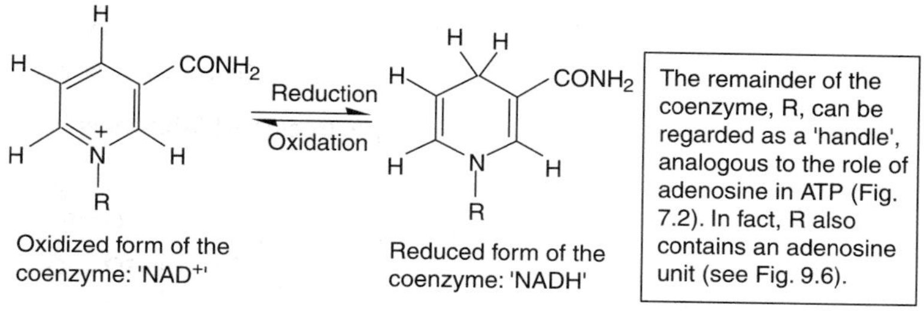

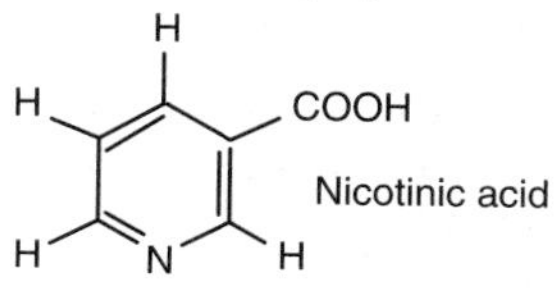

The remainder of the coenzyme, R, can be regarded as a 'handle', analogous to the role of adenosine in ATP (Fig. 7.2). In fact, R also contains an adenosine unit (see Fig. 9.6).

Fig. 7.10 The redox chemistry of nicotinamide coenzymes.

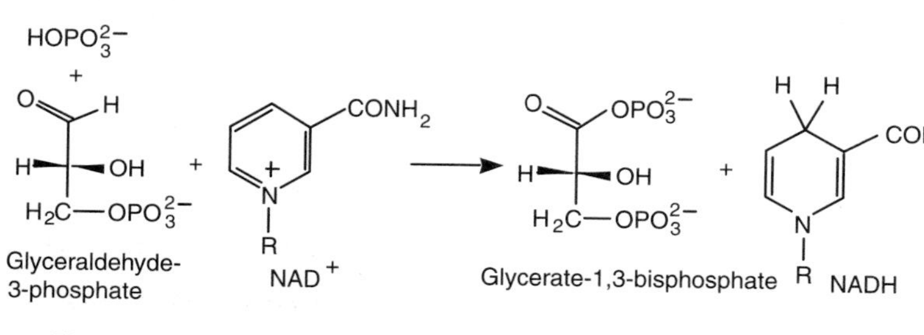

Nicotinic acid

In the formation of glycerate-1,3-bisphosphate, shown in Fig. 7.11, the extra phosphate is introduced from inorganic phosphate, rather than ATP.

Fig. 7.11 Enzyme-catalysed oxidation of glyceraldehyde-3-phosphate.

In contrast to most simple organic species, NADH absorbs light strongly at 340 nm; NAD^+ does not absorb appreciably at this wavelength. Hence the rate of the glyceraldehyde-3-phosphate dehydrogenase reaction can be monitored in a laboratory experiment by following the increase in absorbance at 340 nm as a function of time. This provides a useful assay for the activity of this enzyme, that can also be adapted to measure catalysis by TIM. Glyceraldehyde-3-phosphate (D-GAP) is produced by TIM-catalysed isomerization of dihydroxyacetone phosphate (DHAP). By adding excess glyceraldehyde-3-phosphate dehydrogenase, and appropriate additives, to a mixture of TIM and DHAP, the D-GAP is oxidized as soon as it is formed. Under these conditions, the rate of increase of absorbance at 340 nm is determined by the rate of production of D-GAP, i.e. it is a direct measure of catalysis by TIM. This is an example of a coupled enzyme assay.

At the active site of glyceraldehyde-3-phosphate dehydrogenase, the aldehyde group of the substrate reacts with a thiol group of a cysteine side chain and the resulting intermediate is oxidized by the NAD^+ to produce a thioester. This species undergoes nucleophilic substitution with phosphate to give the acyl phosphate product. Thioesters are important reactive functional groups in biochemistry and they are discussed further in Chapter 8.

The formation of PEP from glycerate-2-phosphate is a rare example of a simple dehydration reaction which is favourable in aqueous solution (equilibrium constant ca 4).

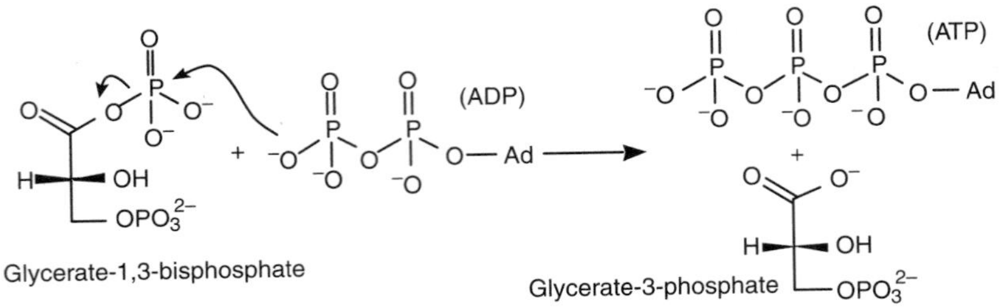

Fig. 7.12 The generation of ATP from glycerate-1,3-bisphosphate.

A further enzyme catalyses the reaction of the resulting acyl phosphate with ADP to generate ATP (Fig. 7.12). In the overall process, chemical energy has been harnessed in the form of a new phosphate anhydride bond. Thus chemical energy, in the form of ATP, has been produced from the overall conversion of glyceraldehyde-3-phosphate to glycerate-3-phosphate.

The product of glycolysis so far is glycerate-3-phosphate. The ATP that was expended in order to accumulate glucose within the cell has been exactly offset by that gained from the oxidation of glyceraldehyde-3-phosphate (two molecules for each glucose metabolized). An overall gain of ATP is obtained by further metabolism of the product. The first two chemical reactions of this sequence produce a molecule capable of producing ATP. The phosphate group is moved along the chain to form glycerate-2-phosphate which then undergoes elimination of water to form phosphoenolpyruvate, PEP (Fig. 7.13).

The phosphate group of PEP is then transferred to ADP to form ATP (Fig. 7.14). In this way two further molecules of ATP are produced from the overall metabolism of each glucose molecule.

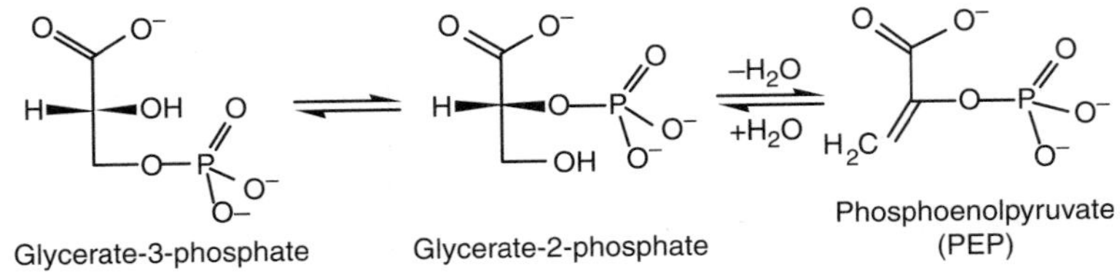

Fig. 7.13 Formation of PEP.

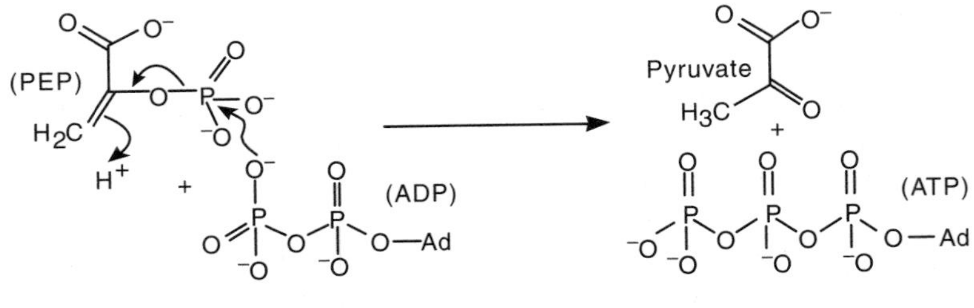

Fig. 7.14 The formation of ATP from PEP.

Further metabolism of the pyruvate and NADH produced by glycolysis provides still more ATP. This process is discussed in Chapter 8.

The phosphorylation of ADP by PEP is favourable because, in the overall reaction, a carbon–carbon double bond in the PEP is replaced by a carbon–oxygen double bond in the final product of the reaction, pyruvate; the latter bond is more stable than the carbon–carbon double bond.

7.5 Pyruvate as a building block

Glycolysis produces two molecules of pyruvate from each molecule of glucose. Pyruvate is a ubiquitous metabolite and undergoes a variety of chemical transformations within cells. As an example, pyruvate has the same carbon skeleton as the amino acid alanine (Fig. 7.15) and cells can interconvert pyruvate and alanine. Many organisms convert pyruvate into alanine, thereby providing one of the amino acids required for protein biosynthesis. This example illustrates the way in which metabolites produced by one set of biochemical transformations can then be used to produce other organic compounds that cells require.

Alanine is one of the amino acids incorporated in proteins (Chapter 2).

Humans have a diet rich in proteins and so they consume plentiful supplies of alanine. Some of this alanine can be converted into pyruvate and metabolized further to produce energy. In this case, the surplus nitrogen liberated is ultimately excreted in the form of urea.

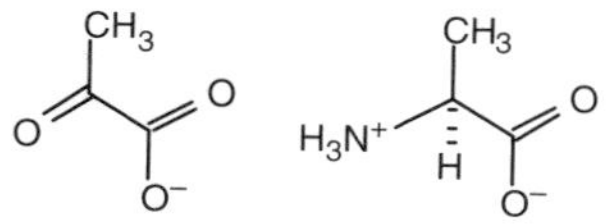

Fig. 7.15 Pyruvate and alanine are related metabolites.

7.6 Summary

Acid anhydrides can act as dehydrating agents, thereby promoting reactions that would normally be unfavourable in water. For this reason, polyphosphates (anhydrides of phosphoric acids) are exploited in biochemistry to store chemical energy. The oxidative metabolism of glucose exemplifies several features of the sugar phosphate chemistry introduced in Chapter 6, including the facile interconversion of sugar derivatives, and the role of anionic phosphate groups in retaining metabolites within cells. It illustrates the way in which complex molecules can be converted into simpler ones with the energy liberated ultimately harnessed in the form of a triphosphate, ATP. These simpler common metabolites can, in turn, be used as building blocks for the biosynthesis of other compounds.

As a further example of the relationship between metabolites produced by glycolysis and amino acids, glycerate-3-phosphate is a biosynthetic precursor of serine and cysteine.

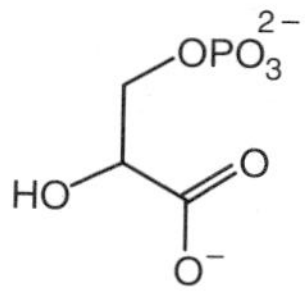

Glycerate-3-phosphate

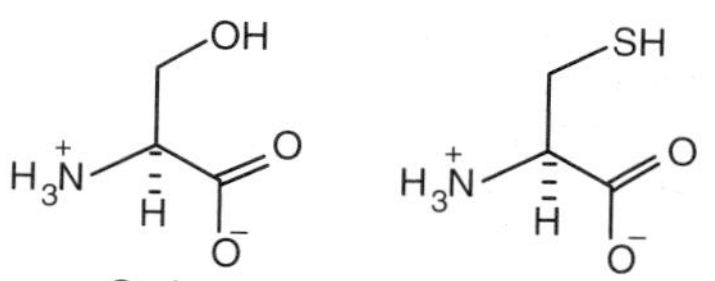

Serine   Cysteine

The oxidation reactions of sugar metabolism require chemical reactivity not available in simple proteins. Coenzymes are conscripted to provide the needed reactivity. A range of coenzymes are exploited in biochemistry for specific catalytic tasks. Coenzymes typically consist of a functional group, with the distinctive chemistry required, attached to a 'handle', that facilitates effective binding by enzymes. They are discussed further in Chapter 9.

Thioesters are used alongside polyphosphates as biological dehydrating agents. They have been mentioned in passing in this chapter (see margin note in Section 7.4.3) and will be discussed further in Chapter 8.

Further reading

All major biochemistry textbooks, such as those referred to at the end of Chapter 1, discuss glycolysis. Chemical energetics and glucose metabolism is particularly well covered in R. H. Abeles, P. A. Frey and W. P. Jencks (1992) *Biochemistry*, Jones and Bartlett Publishers Inc., chapters 9 and 21.

The chemistry of glycolysis is discussed in chapter 25 of D. S. Kemp and F. Vellacio (1980) *Organic Chemistry*, Worth Publishers Inc., NY.

Phosphate biochemistry is discussed by F. H. Westheimer (1987) Why Nature Chose Phosphates, *Science*, **235**, pp. 1173–8.

8 Lipids: cells as compartments

8.1 Introduction

Compartmentalisation is an essential part of life as we know it; the contents of cells are different from their surroundings. Cells maintain a distinct repertoire of chemical processes and in order to do so they minimise the uptake of undesirable chemicals and the loss of desirable ones. The compartments of living organisms are bounded by semipermeable membranes. So far, attention has been focused on the internal contents of cells, a collection of water-soluble chemicals. This chapter examines the molecular components of the membranes enclosing these compounds.

Most organic compounds are immiscible with water and tend to associate in aqueous solution. This association is exploited by cells, which use a class of compounds, the lipids, to form the basic structures of cell membranes.

This chapter develops an awareness of the intrinsic chemistry of the most common lipids, phospholipids; the basis of their association to form membranes; and some of the implications of compartmentalisation.

As discussed in Chapter 3, proteins in aqueous solution tend to adopt structures in which polar residues, which interact favourably with solvent, lie predominantly on the surface; non-polar residues tend to be buried away from water. The factors responsible for the association of lipids into membranes are closely related to those that determine the tertiary structures of proteins.

Amphiphilicity was introduced in Section 3.3, with reference to α-helices. Amphiphilic helices are often found in soluble protein structures; the hydrophilic face of the helix hydrogen bonds with water and points towards the solution, whilst the hydrophobic face is buried away from water.

8.2 Common phospholipids found in membranes

The membranes found in biological systems are formed by the spontaneous association of relatively small organic molecules, lipids (Fig. 8.1). The membrane structures arise because of the amphiphilic nature of lipid molecules; one end (the 'polar head group') is hydrophilic, the remainder is a hydrophobic tail.

The most common lipids, phospholipids, are sugar phosphate derivatives. Glycerol-3-phosphate (Fig. 8.2), the core of phospholipids, is ideally suited to act as a molecular scaffold. Both the acid group of the phosphate

The biochemistry of sugar phosphates was introduced in Sections 6.5 and 6.6.

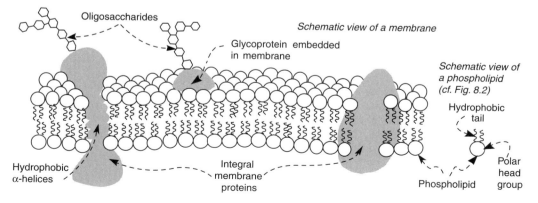

Fig. 8.1 A typical membrane and its constituent lipids.

Glycerol-3-phosphate, the sugar phosphate core of phospholipids, is simply a reduced form of the triose phosphates, e.g. DHAP, which have been discussed in Chapters 5–7.

Phosphate, and its monoesters, exist as a mixture of mono- and di-anions at neutral pH. For clarity in this chapter such phosphate derivatives are represented as mono-anions.

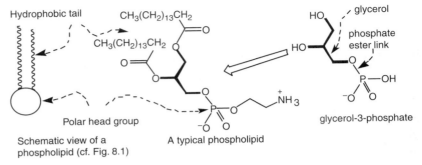

Schematic view of a phospholipid (cf. Fig. 8.1)

A typical phospholipid

glycerol-3-phosphate

Fig. 8.2 The core of phospholipids is derived from glycerol-3-phosphate.

and the hydroxyl groups of the sugar derivative are useful sites for the attachment of other functionalities. The ability of both sugar and phosphate portions to form esters allows sugar phosphates to act as linking groups. In phospholipids, the polar phosphate group combines with polar alcohols to form the hydrophilic head group, and large non-polar chains are connected to the sugar portion to produce the hydrophobic tail.

8.2.1 The hydrophobic tail of phospholipids

In mammals, the chains of fatty acids incorporated into lipids are typically between 12 and 20 carbons in length. They comprise an even number of carbons because they are biosynthesized by successive additions of a two-carbon unit—a derivative of acetic acid, acetyl CoA (the generation of acetyl CoA is mentioned in an aside to Section 8.6). Common unsaturated fatty acids contain up to six double bonds; the double-bond geometry is usually *cis* and, where more than one double bond is found, there is usually a single methylene group separating the double bonds.

In phospholipids, the two free hydroxyl groups of glycerol-3-phosphate form ester links to a class of carboxylic acids—fatty acids. Fatty acids consist of a linear hydrocarbon chain with a carboxylic acid group at one

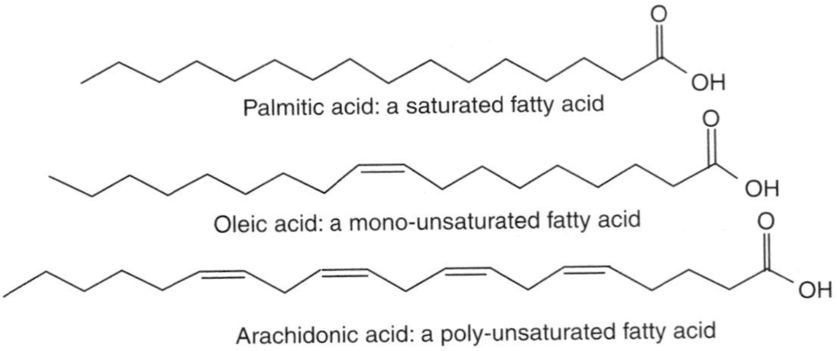

Palmitic acid: a saturated fatty acid

Oleic acid: a mono-unsaturated fatty acid

Arachidonic acid: a poly-unsaturated fatty acid

Fig. 8.3 Typical fatty acids.

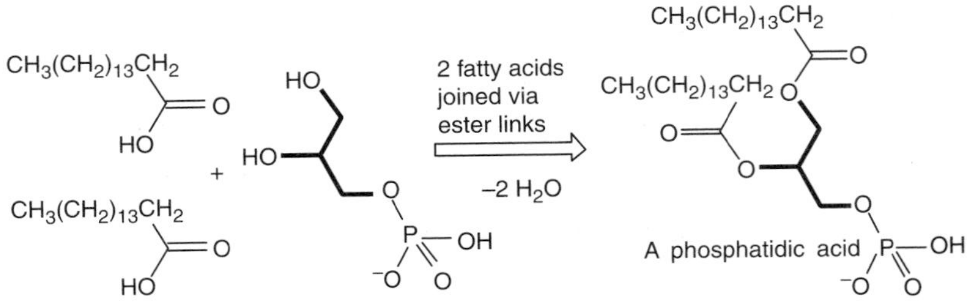

Some cells generate esters of glycerol with three links to fatty acids. These triglycerides, found for example in animal fat, are insoluble in water. They are not used in membranes but are deposited inside cells, providing stores of useful chemicals; they are laid down in times of plenty, and metabolized in times of need to provide energy.

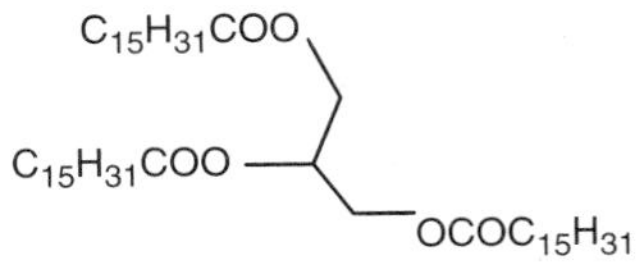

A triglyceride, tripalmitoyl glycerol

2 fatty acids joined via ester links

$-2 H_2O$

A phosphatidic acid

Fig. 8.4 Phosphatidic acids: esters derived from fatty acids and glycerol-3-phosphate.

end. Many fatty acids are devoid of other functionalities; some, however, contain double bonds. Three common fatty acids are shown in Fig. 8.3.

The combination of glycerol-3-phosphate and two fatty acids is known as a phosphatidic acid (Fig. 8.4); such compounds are typical phospholipids.

8.2.2 The polar head group of phospholipids

Phosphatidic acids are phosphate monoesters capable of forming further ester linkages. In the common phospholipids, the diacyl-glycerol-3-phosphate is linked to an alcohol bearing one or two charged groups. This arrangement provides a highly hydrophilic head group. Serine, one of the amino acids found in proteins, is one such alcohol. The other alcohols commonly utilised, ethanolamine and choline, are related to serine.

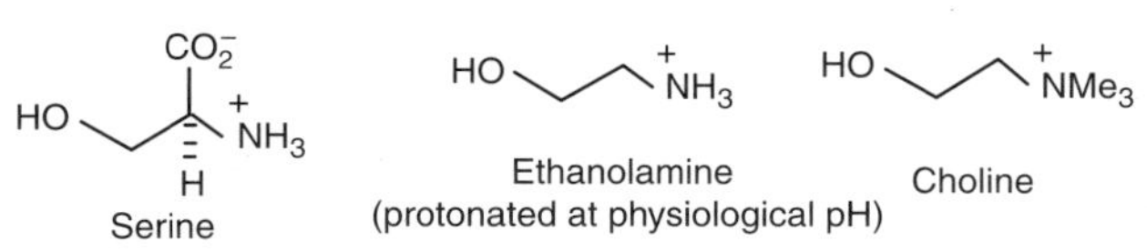

Ethanolamine is derived from serine by decarboxylation (loss of carbon dioxide). Addition of three methyl groups to the amino group of ethanolamine results in choline.

The most common phospholipids, phosphatidylethanolamine and phosphatidylcholine, are phosphate diesters (Fig. 8.5) in which the phosphate group retains a negative charge, whilst the polar alcohol introduces a positively charged functional group. The presence of a pair of matched charged groups at one end of a phospholipid augments the polarity difference within the lipid molecule but leaves these lipids with no net overall charge, they are 'Zwitterions'.

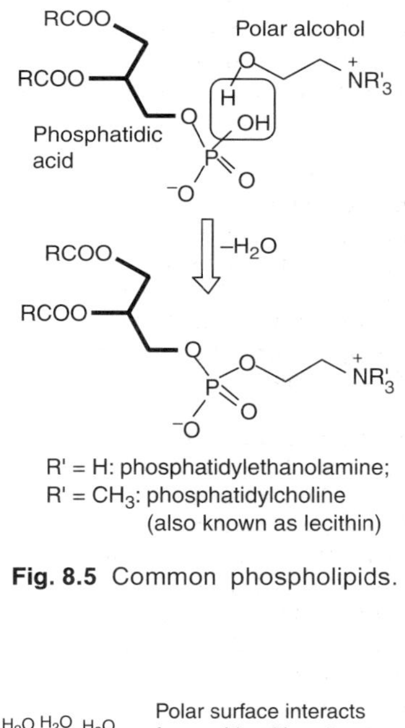

8.3 Biological membranes: association of lipids

When phospholipids, such as phosphatidylethanolamine, are introduced into water, the charged hydrophilic 'head group' interacts favourably with water. By contrast, the large non-polar surfaces of the fatty acid chains disrupt the hydrogen bonding properties of water. To avoid this unfavourable interaction with water, lipid molecules associate to form large assemblies that prevent the fatty acids from coming into contact with water (Fig. 8.6).

The overall structure adopted is like a sandwich, with the polar head groups acting as slices of bread on the outside, and the non-polar groups buried as a filling. A layer of lipid molecules presents an array of polar

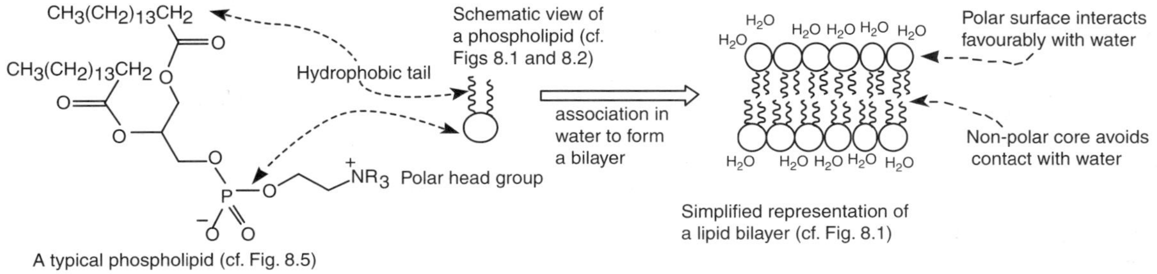

R' = H: phosphatidylethanolamine;
R' = CH₃: phosphatidylcholine (also known as lecithin)

Fig. 8.5 Common phospholipids.

Fig. 8.6 The association of lipids to form a bilayer.

The proteins found in membranes are specifically inserted there for functional purposes. They expose largely non-polar surfaces to the non-polar core of the membrane. This situation contrasts with water-soluble proteins, where such residues are found buried away from the surface. Membrane-bound proteins can often be identified from their amino acid composition: the regions of the molecule embedded in a membrane are rich in non-polar residues.

Proteins and oligosaccharides provide a wide variety of molecular structures. When membrane bound, they play an important role in molecular recognition processes. For example, antibodies are glycoproteins (proteins with oligosaccharides attached) found on the surface of some blood cells. They recognise possible infective agents by interacting specifically with foreign molecules; these molecular recognition processes mediate the immune response.

Proteins spanning the cell membrane allow communication with the outside world. Some play a role in 'signal transduction' where specific interactions with a molecule outside the cell trigger changes inside the cell. Hormones, e.g. insulin, present in the bloodstream exert their influence on the activity of cells in this way.

Glucose metabolism provides an example of the concentration of chemicals from the environment. As was discussed in Chapter 7, the first step in glycolysis is phosphorylation of glucose. This step consumes ATP and is used to drive the sequestration of glucose from the surrounding media.

functional groups to the solvent. The corresponding non-polar chains associate with the non-polar surface of a second layer of lipids. In associating to form a bilayer, the hydrophobic portions of lipids become buried away from water. The core of the bilayer is a sheet of non-polar material that prevents the free passage of polar species from one side to the other.

8.4 Cells as compartments

Cells are surrounded by a membrane. This allows them to maintain a controlled environment appropriate for their chemical processes. Cells are, however, in dynamic interaction with the world, since the membrane is semipermeable and allows cells to concentrate useful chemicals, such as foodstuffs. There is also a selective advantage for cells to deplete the levels of some chemicals relative to their environment, e.g. it is important to excrete undesirable materials. The movement of chemicals into and out of cells is mediated, primarily, by proteins embedded within the membrane. The membranes surrounding cells are thus rather more complex than simple lipid bilayers (Fig. 8.7).

Proteins are more highly functionalised than lipids and add chemical variety to the membrane. In particular, many membrane-bound proteins allow the migration of selected species through the membrane. Some of these channels simply allow free migration of a particular ion or molecule through the membrane, thereby equilibrating internal and external concentrations. Others act as pumps, or as gates—opening selectively in response to specific needs of the cell.

The concentration of nutrients is of obvious benefit to cells. Cells use energy to concentrate some chemicals internally, thereby creating concentration gradients with the environment. Cells also actively pump out some chemicals. This is not just true of unwanted waste products; it is also true of some simple ions, such as protons and metal ions. Once the concentration of ions is higher outside the cell than inside, there is a thermodynamic driving force for the ion to return and equalise the concentrations. This pumping mechanism is actually exploited as a store of chemical energy.

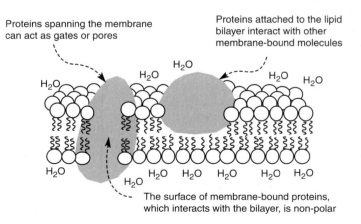

Fig. 8.7 A schematic view of membrane-bound proteins in a lipid bilayer.

Proteins in the membrane open and close to act
as selective pumps and gates to control ion flow

Z^+ Z^+ Z^+ Z^+ Z^+

outside cell

Energy is expended, actively
pumping out Z^+, to create a
concentration imbalance

Energy is liberated when Z^+ is
allowed to flow back into the cell
to equalize the concentrations

inside cell

Z^+ Z^+ Z^+ Z^+ Z^+

*Low-energy situation: equal concentrations of Z^+ inside
and outside the cell; no driving force for ions to flow*

Z^+ Z^+ Z^+ Z^+ Z^+ Z^+ Z^+ Z^+

outside cell

inside cell

Z^+

*High-energy situation: the concentration imbalance
makes it favourable for Z^+ ions to flow into the cell*

Fig. 8.8 Schematic view of the manipulation of concentration gradients by cells, illustrated for a typical ion Z^+.

Table 8.1 Approximate concentrations of some common ionic species (mmol L^{-1})

	Ca^{2+}	Cl^-	Na^+	K^+
Sea water	10	550	450	10
Squid nerve cell (outside)	10	550	450	10
Squid nerve cell (inside)	1	100	50	400
Mammalian muscle cell (outside)	2	120	150	5
Mammalian muscle cell (inside)	10^{-7}	5	10	150

The importance of inorganic ions for living systems was discussed in Chapter 1.

Sodium–potassium ATPase hydrolyses ATP to create concentration gradients of ions. The reverse process is also possible, namely proteins can synthesize ATP at the expense of dissipating a concentration gradient. An important example of this is described in Section 8.5.

Favourable chemical reactions are used to establish a concentration gradient. When ions are allowed to re-enter the cell, to redress the imbalance, the flow of ions is coupled to the regeneration of chemical energy (Fig. 8.8).

Sea water is rich in sodium chloride, together with other ions such as potassium, magnesium and calcium. These ions are essential for all life. Cells manipulate the relative concentrations of these ions inside and outside cells, and exploit the resulting concentration gradients in a variety of ways. As an example, it is common for multicellular organisms to maintain comparatively low concentrations of sodium ions within cells relative to the extracellular environment (Table 8.1). In animals this is achieved by the action of a membrane-bound protein, sodium–potassium ATPase. This protein is a pump that exploits the energetically favourable hydrolysis of ATP to move sodium ions out of the cell and potassium ions into the cell. For each ATP molecule that is hydrolysed, three sodium ions are pumped out of the cell and two potassium ions are pumped in. The action of this pump generates concentration gradients of these ions and leads to an electrical potential across the cell membrane. Such phenomena are exploited by cells for a variety of purposes.

With a much higher concentration outside the cell than in, the re-entry of sodium ions is favourable. As an example, some cells in the intestine use this driving force to facilitate the uptake of nutrients (Fig. 8.9). Special

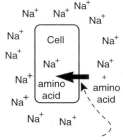

Pore in membrane allows favourable return of Na^+, only if accompanied by an amino acid

Fig. 8.9 Schematic view of the coupling of sodium ion gradients to amino acid uptake.

proteins act as channels in the membrane allowing sodium ions to enter the cell, but only if this process is accompanied by the entry of an amino acid, e.g. alanine, or a monosaccharide, e.g. glucose. In this way the favourable entry of sodium ions drives the uptake of another nutrient.

The propagation of electrical signals along nerve cells is a related phenomenon. In response to a chemical signal from a neighbouring nerve cell, sodium ions (Table 8.1) are allowed to flow into a nerve cell through sodium ion-selective pores. This inflow of cations increases the positive charge within the cell and hence results in a change in the electrical potential across the membrane. The result is that an electrical signal propagates along the nerve cell to transmit the nerve signal.

8.5 Oxidative phosphorylation

Hydrogen is a central element for life, being a key component of water and organic compounds. In addition, concentration gradients of hydrogen ions, protons, are a critical feature of cellular activity; they are used to generate ATP as part of a two-stage process termed 'oxidative phosphorylation' (Fig. 8.11). Favourable redox reactions are carried out at a membrane and used to drive the export of protons, thereby producing a proton concentration gradient. In a second stage, the re-entry of protons is coupled to the production of ATP. This coupling is discussed in a little more detail below.

The oxidative conversion of glucose to pyruvate (Chapter 7) produces two molecules of ATP directly for each molecule of glucose that has been metabolised. The other product of the redox chemistry is the reduced nicotinamide compound, NADH. In the presence of an appropriate oxidising agent, this species can be reoxidised. The oxygen in an aerobic atmosphere is a powerful electron acceptor. Electrons liberated from NADH are transferred indirectly to oxygen, reducing it to water (Fig. 8.10). This series of favourable electron transfer steps take place at the cell membrane and is coupled to the export of protons (Fig. 8.11). The favourable return of protons, which redresses the concentration imbalance, is used to generate ATP.

In eukaryotic cells (cells with a nucleus, such as those found in animals), oxidative phosphorylation (Fig. 8.11) is carried out in specialised bodies

Metal ions, other than sodium, are also used in biological signalling. For example, the concentration of calcium ions in the cytosol of eukaryotic cells is kept low (Table 8.1). The selective uptake of calcium ions provides a signalling mechanism for a wide range of processes, such as the activation of enzymes involved in blood clotting and digestion.

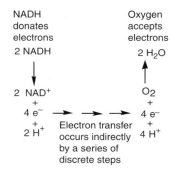

Fig. 8.10 Oxidation of NADH to NAD$^+$, and reduction of O$_2$ to H$_2$O occur in parallel at membranes.

In anaerobic organisms, electron acceptors other than oxygen are used for this form of energy generation. For example, some bacteria use carbon dioxide to accept electrons. The carbon dioxide ends up as methane. These organisms, methanogens, are responsible for the methane produced in marshes (marsh gas) and in the insides of ruminants such as cows.

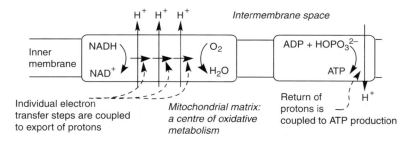

Fig. 8.11 Schematic view of oxidative phosphorylation by mitochondria.

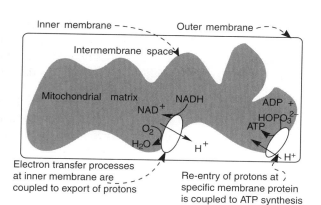

Inner membrane

Outer membrane

Intermembrane space

Mitochondrial matrix

NAD$^+$ NADH

ADP +
HOPO$_3^{2-}$
ATP

O$_2$
H$_2$O   H$^+$

H$^+$

Electron transfer processes
at inner membrane are
coupled to export of protons

Re-entry of protons at
specific membrane protein
is coupled to ATP synthesis

Fig. 8.12 Overview of oxidative phosphorylation at mitochondria.

Mitochondria are small bodies ('organelles') capable of carrying out oxidative phosphorylation; they are found within eukaryotic (but not prokaryotic) cells. By comparison, aerobic bacteria undertake oxidative phosphorylation at the bacterial membrane. The mitochondrial and bacterial systems are very similar. It is thought that they are ancestrally related. On this view, mitochondria are descendents of bacteria that established an interdependent ('symbiotic') relationship with early eukaryotic cells.

within the cell, known as mitochondria (Fig. 8.12). A series of electron transfer steps is catalysed by protein complexes embedded in the inner membrane. This electron transfer chemistry results in the export of protons from the interior of the mitochondria to an intermembrane space. A further protein complex in the inner membrane acts as a channel to allow the re-entry of protons but only when coupled to the synthesis of ATP from ADP and phosphate.

Oxidative phosphorylation liberates the equivalent of three molecules of ATP for each NADH molecule reacted. This corresponds to six molecules of ATP for each molecule of glucose undergoing glycolysis. Therefore, the majority of ATP produced by glycolysis arises from the indirect means of oxidative phosphorylation rather than directly as described in Chapter 7.

The protein complex that mediates proton re-entry and ATP production, F_1–F_0 ATPase, mediates a similar type of process to the Na–K ATPase (Section 8.4). In this case the process is run in the reverse direction to produce ATP from a concentration gradient. F_1–F_0 ATPase is a complex assembly of proteins which acts like a molecular motor. It is a good example of a molecular device (see Section 1.6).

8.6 Summary

Sugars and phosphates can both form esters with more than one partner, and so are useful as linking units in the construction of complex molecules. A phosphate group retains its polar ionic character when linked, in this fashion, to two alcohols. By appending a combination of non-polar fatty acid groups and a polar alcohol to a glycerol phosphate core, molecules containing both non-polar and polar regions are produced. These lipid molecules spontaneously associate in water to bury the non-polar regions, whilst leaving the polar groups exposed. As a result they form membranes which are exploited by cells as a means of compartmentalisation. Proteins embedded within the membrane add chemical diversity and render the membrane semipermeable; they interact with the outside world, and can be used to manipulate the flow of chemicals into and out of cells. Compartmentalisation allows the generation and exploitation of concentration gradients of ions. These gradients play an important role in many aspects of biochemistry, such as signalling and energy generation.

As well as oxidising the reduced cofactors produced by glycolysis, mitochondria also mediate the further oxidative metabolism of pyruvate. Firstly pyruvate is oxidised to produce carbon dioxide and a thioester derivative of acetate, acetyl CoA. The latter is then oxidised by a series of reactions known as the tricarboxylic acid cycle (also known as the citric acid cycle, or the Krebs cycle). The overall effect of this cycle is to convert the two carbons of acetyl CoA into carbon dioxide and generate more nucleoside triphosphate, both directly and via further oxidative phosphorylation of the reduced cofactors that are produced. A combination of all these processes produces 38 molecules of ATP for each molecule of glucose oxidized completely to carbon dioxide.

It is a useful exercise to analyse the chemistry of the oxidative metabolism of pyruvate and relate it to the metabolic chemistry described in Chapters 7 and 8.

Further reading

Major biochemistry textbooks, such as those listed in Chapter 1, all discuss lipids and compartmentalisation. An accessible text on bioenergetics is F. M. Harold (1986) *The Vital Force: A Study of Bioenergetics*, W. H. Freeman and Co., New York.

A detailed account of membrane biochemistry is given in R. B. Gennis (1989) *Biomembranes: Molecular Structure and Function*, Springer-Verlag.

9 Genetic information: nucleotides and nucleic acids

9.1 Introduction

Cells are complex entities, localized capsules of chemicals that can reproduce to generate new cells of a near-identical nature. The nature of a particular cell depends on the array of chemicals within the cell and the nature of their interconversion. Proteins play a key role in establishing this identity; they are intimately involved in all the activities of the cell, e.g. the uptake and transport of small molecules, and the catalysis of chemical reactions. The information used by a cell is, therefore, related to the repertoire of proteins present.

In general, a cell contains a 'permanent' information store that contains the blueprint needed to carry out all the processes of which the cell is capable. This is the genetic information that is inherited when a cell divides and produces more cells. Not all of this information will be in use at any one time and place. Cells of all kinds make and use particular proteins at different times, in response to particular demands; this flexibility derives from the use of a 'transient' form of information.

'Permanent' and 'transient' information storage in cells is associated with nucleic acids: the polymers *deoxyribonucleic acid* (DNA) and *ribonucleic acid* (RNA), respectively. These polymers are built from nucleotide monomer units. The information carried by nucleic acids is contained in the nature of the monomer units and their order along the polymeric chain. Amongst other things, this ordering provides the information for the construction of the proteins produced by cells.

This chapter will discuss the nature of nucleotides and of the polymers that derive from them—nucleic acids. These polymers fold into ordered structures for similar reasons to those that result in protein folding and give rise to the spontaneous assembly of lipids into membranes. The structural features of nucleic acids, which are exploited in their role as information stores, are described first. This is followed by an outline of the way in which this information is accessed and transmitted from generation to generation.

The use of different information by cells with equivalent genetic inheritance is a feature of both multicellular and unicellular creatures. For example, human brain cells and liver cells possess the same 'permanent' information, but whilst each uses some proteins in common, some specialized proteins are produced only in liver cells and some only in brain cells.

Similarly, microorganisms tailor the range of enzymes they make to optimize the use of the nutrients available in their surroundings.

Regulating the activity of proteins is a key feature of cells. Control of the rate of protein synthesis is known as the regulation of gene expression (see Section 9.11). In addition, the activity of many individual proteins within a cell can be modulated; this is often achieved by the allosteric properties of multimeric proteins (e.g. haemoglobin, as discussed in Chapter 4).

9.2 Nucleotides

Nucleotides are formed from three components: a sugar, ribose or 2-deoxyribose; a phosphate group; and a heterocyclic base (Fig. 9.1), one of a set of related, nitrogen-containing compounds.

Heterocyclic compounds were introduced in Section 2.3. In the case of heterocyclic bases, nitrogen atoms are present in the rings.

Nucleotides based on ribose are used for a wide variety of tasks in biochemistry—both as small molecules, e.g. ATP, and incorporated into the polymer, RNA. By contrast, the biochemistry of nucleotides derived from 2-deoxyribose is far more specialized; it is dominated by their use as monomers for the construction of DNA. The choice of 2-deoxyribose is closely related to the specific role of DNA as a long-term information store (see Section 9.10).

Three of the heterocyclic bases, adenine, guanine and cytosine, are common to both RNA and DNA. RNA and DNA differ, however, in the fourth base employed: uracil is found in RNA and thymine is found in DNA. The difference in structure is only subtle: a methyl group replaces one of the hydrogen atoms attached to the ring. This has no significant effect on the chemical properties, and the two bases behave in very similar ways.

In the acyclic form of ribose, the carbonyl group is an aldehyde; ribose is, therefore, an aldo-pentose. The relationship between the acyclic and five-membered ring forms of ribose is shown in Fig. 6.5. In the cyclic structures of ribose and 2-deoxyribose, the former carbonyl carbon, C-1, is still distinct and recognizable by the fact that it forms two bonds to oxygen.

Although RNA is built from four basic monomers, modification of some bases is observed in certain circumstances. This parallels the situation in proteins where a basic repertoire of 20 building blocks is sometimes augmented by modification of a side chain after biosynthesis of a protein molecule.

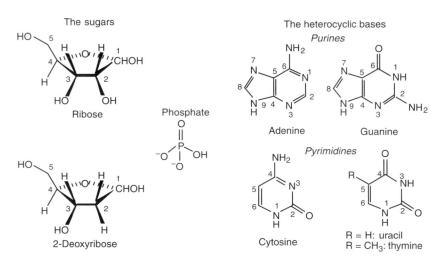

Fig. 9.1 The components of nucleotides and nucleic acids.

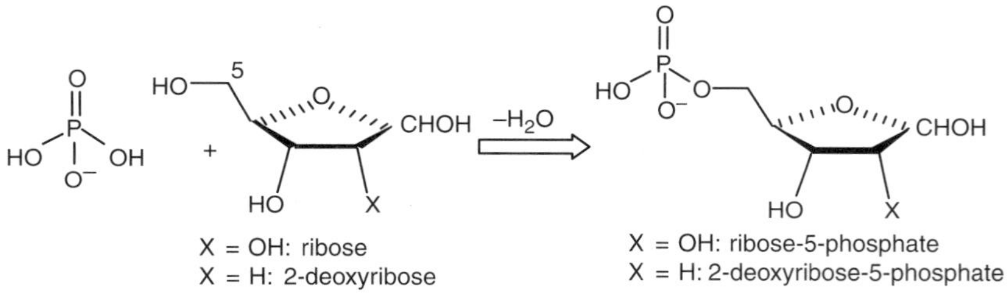

Fig. 9.2 The sugar phosphate backbone of common nucleotides.

9.2.1 The sugar phosphates of nucleotides

The sugar found in most nucleotide derivatives, including RNA, is a five-carbon sugar, ribose. DNA is based on a closely related sugar, 2-deoxyribose, a ribose derivative in which the hydroxyl group on C-2 is replaced by hydrogen. As explained in Chapter 6, these sugars adopt cyclic structures.

Like a lipid, a nucleotide is a phosphate ester of a sugar. Ribose and 2-deoxyribose both have more than one free alcohol group and so can form more than one ester link. The most common unit for nucleotides, including that forming the building blocks of DNA and RNA, involves a phosphate ester linkage to the primary alcohol function at C-5 (Fig. 9.2).

9.2.2 Heterocyclic bases: the third component of nucleic acids

The component that distinguishes a nucleotide from the other classes of molecule discussed so far is the heterocyclic base. The bases found in nucleotides are planar, cyclic, aromatic molecules containing an array of nitrogen-based functional groups; oxygen-derived functional groups are also present in three of the bases, and, like the nitrogen-based ones, are capable of hydrogen bonding.

There are two classes of heterocyclic bases (Fig. 9.1); both contain a six-membered ring with two nitrogens in a 1,3 arrangement. In one class, only

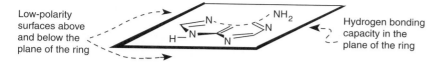

Fig. 9.3 Adenine illustrates the hydrogen bonding and low-polarity features of bases.

a single ring is present; these heterocycles are known as pyrimidines. In the other class, a second (five-membered) ring, an imidazole, is fused onto the six-membered ring, resulting in a bicyclic structure known as a purine. The biologically important purines and pyrimidines are each of two types, depending on whether they have oxygen or nitrogen substituents at key positions on the ring (see Fig. 9.1).

The bases as a whole are not, however, very polar molecules, and they are not readily solvated by water: the faces of the rings provide a large surface area of low polarity (see Fig. 9.3); the edges provide hydrogen bonding capability.

9.2.3 The structures of nucleotides

Nucleotides are generated by forming a link between a particular nitrogen atom of a heterocyclic base (N-1 of pyrimidines and N-9 of purines) and the C-1 atom of the sugar phosphate unit (Fig. 9.4).

There are four types of nucleotide in RNA and DNA. They are distinguished by the identity of the particular base incorporated (Fig. 9.5). In the case of ribonucleotides, the four bases are adenine, guanine, cytosine and uracil. The combination of a heterocyclic base and ribose, without a phosphate group present, is known as a nucleoside: adenosine, guanosine, cytidine and uridine, respectively. Nucleotides are generally named as phosphate esters of nucleosides. Hence the corresponding nucleotides are adenosine-5'-monophosphate (AMP), guanosine-5'-monophosphate (GMP), cytidine-5'-monophosphate (CMP) and uridine-5'-monophosphate (UMP). The presence of the deoxyribose sugar is sometimes denoted by the prefix d. Hence, the abbreviated names for the four deoxyribonucleoside-5'-monophosphates are dAMP, dGMP, dCMP and dTMP (Fig. 9.5).

An imidazole group is also found in the side chain of the amino acid histidine, see Section 2.3.

The links between the components of nucleotides correspond to the products of dehydration reactions. The latter are all ultimately driven by the chemical reactivity of polyphosphates, e.g. ATP. The bond-forming processes involve nucleophilic substitution reactions with phosphates utilized as good leaving groups.

In the structure of nucleotides the heterocyclic base is found in a particular orientation relative to the sugar ring. It is above the plane of the ring, as nucleotides are conventionally drawn. This is the so-called β-anomer (see Section 6.3).

The full names of the nucleotide building blocks of DNA are 2'-deoxyadenosine-5'-monophosphate (dAMP), 2'-deoxy-guanosine-5'-monophosphate (dGMP), 2'-deoxycytidine-5'-monophosphate (dCMP), and 2'-deoxythymidine-5'-monophosphate (dTMP).

Care must be taken to avoid confusion with the names of heterocyclic bases and nucleosides. For example, ade<u>nine</u> is a base and ade<u>nosine</u> a nucleoside, but cyto<u>sine</u> is a base and cyt<u>idine</u> a nucleoside.

Fig. 9.4 A schematic view of the link between base and sugar phosphate in nucleotides.

A complication arises in numbering the positions of atoms in nucleotides since there are two discrete frameworks: the heterocyclic base and the ribose ring. To avoid ambiguity, the atoms of the base are numbered normally and those of the ribose distinguished by the use of a prime, i.e. $1'$, $2'$, $3'$, $4'$ and $5'$.

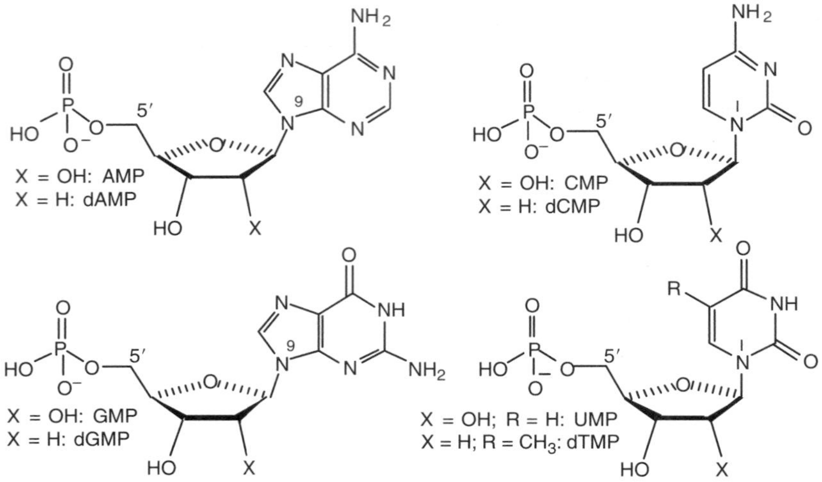

X = OH: AMP
X = H: dAMP

X = OH: CMP
X = H: dCMP

X = OH: GMP
X = H: dGMP

X = OH; R = H: UMP
X = H; R = CH_3: dTMP

Fig. 9.5 Nucleotide structures.

9.3 Some important nucleotide derivatives

Remember that a coenzyme is the name given to an organic molecule conscripted by an enzyme to aid in catalysis of a particular chemical transformation.

Nucleotides are important components of all cells; they are used both for their intrinsic chemistry and as building blocks. AMP, in particular, provides a molecular scaffold for several coenzymes. The coenzymes NAD^+ and CoA-SH, encountered in Chapters 7 and 8, consist of a reactive functional group (the basis of the chemical role of the coenzyme) appended to a nucleotide group that includes an AMP unit (Fig. 9.6).

ATP is a nucleotide bearing a triphosphate, rather than a monophosphate, group at the $5'$-position. It can be viewed as a derivative of AMP. As discussed in Chapters 7 and 8, triphosphates are utilized as a form of chemical energy in cells.

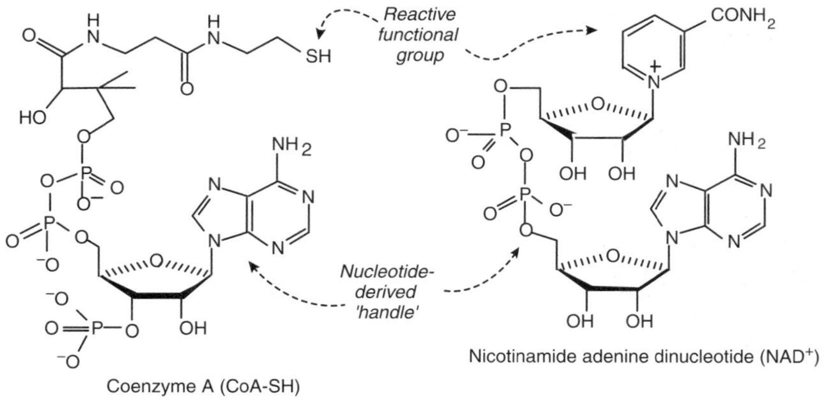

Reactive functional group

Nucleotide-derived 'handle'

Coenzyme A (CoA-SH)

Nicotinamide adenine dinucleotide (NAD^+)

Fig. 9.6 CoA-SH and NAD^+: coenzymes built on a nucleotide framework.

9.4 Nucleic acids are polymers of nucleotides

Details of the biosynthesis of nucleic acids are discussed in Section 9.11.

Nucleotides are phosphate monoesters and, as such, can form further ester links to other alcohols. Since nucleotides themselves possess further alcohol groups, polymers can be formed by dehydration chemistry between nucleotides. Nucleic acids are polymers of this type, in which each sugar phosphate is linked to the hydroxyl group on the C-$3'$ of another ribose phosphate unit. The oxygen substituents involved lie on opposite faces of the sugar, and this phosphodiester arrangement leads to an extended linear polymer that bears a single anionic charge at each phosphate centre (Figs 9.7 and 9.8).

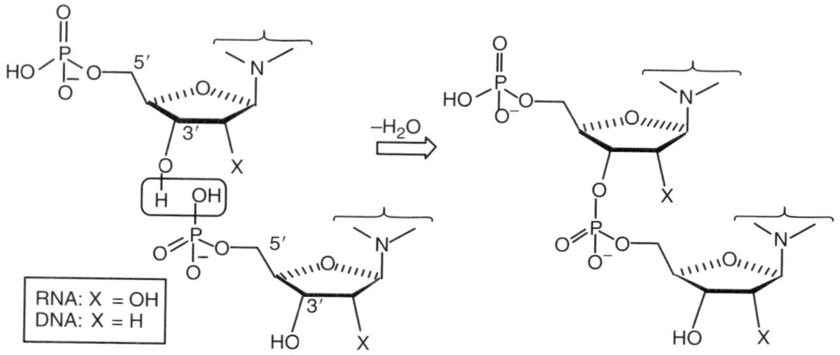

Fig. 9.7 The phosphodiester link of nucleic acids.

9.5 Primary structure of nucleic acids

In nucleic acids the individual nucleotides are identified by single-letter abbreviations. The order of monomer residues along the polymer chain can therefore be designated by a string of letters (Fig. 9.8), provided that the directionality of the chain is clear. Conventionally, nucleic acid structures are represented as sequences written in the 5' to 3' direction.

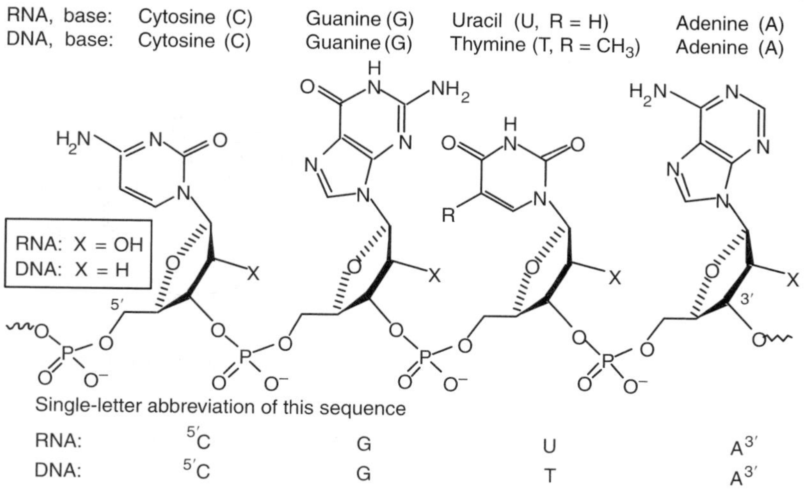

The existence of directionality in nucleic acid chains is analogous to the situation in the polypeptide chains of proteins (Chapter 3). The convention of writing nucleic acid sequences in the 5' to 3' order is akin to describing protein sequences in terms of the appearance of residues in the N to

In accordance with this convention, the directionality of the nucleic acid chain depicted in Fig. 9.8 is

5' ⟹ 3'

RNA, base:	Cytosine (C)	Guanine (G)	Uracil (U, R = H)	Adenine (A)
DNA, base:	Cytosine (C)	Guanine (G)	Thymine (T, R = CH$_3$)	Adenine (A)

RNA: X = OH
DNA: X = H

Single-letter abbreviation of this sequence

RNA:	$^{5'}$C	G	U	A$^{3'}$
DNA:	$^{5'}$C	G	T	A$^{3'}$

Fig. 9.8 Representations of a specimen nucleic acid sequence.

9.6 RNA folds into well-defined shapes

RNA carries out a range of biochemical tasks. There are three major classes of RNA molecules in cells, categorized according to their different

As is discussed later, either DNA or RNA can act as the repository of genetic information for viruses.

Ribosomes are particles made up of RNA and protein, found in the cytoplasm of cells. They are the catalytic machinery through which proteins are produced.

The way in which the amphiphilic character of lipids leads to ordered bilayer structures was discussed in Chapter 8.

As was described in Chapter 3, in any non-planar polymer where succeeding monomer units adopt the same conformations the overall structure is a helix.

As was introduced in Chapter 7, tRNA forms esters with amino acids and these are used in the biosynthesis of proteins. tRNA plays the role of an adaptor molecule, translating the information of nucleic acids into that of proteins. There is at least one tRNA molecule for each amino acid incorporated into proteins. tRNA^{Phe} refers to a tRNA molecule dedicated to the introduction of phenylalanine (Phe) into proteins.

This topic is described in more detail in Section 9.11; some of the labels in Fig. 9.9 refer to issues discussed there.

biochemical function: messenger RNA (mRNA), which carries the information for the sequence of a protein from DNA; transfer RNA (tRNA), which decodes this information during the biosynthesis of proteins; and ribosomal RNA (rRNA), which is associated with ribosomes, the protein-synthesizing machinery. By contrast, DNA has a single biochemical role, as the molecule associated with storage of genetic information. In the following discussion the structural properties of RNA, which make it such a versatile polymer, will initially be elaborated. The structural features of DNA, which equip it for its specialized biochemical niche, will be described subsequently.

Like proteins and lipids, nucleic acids are amphiphilic molecules. These molecules adopt ordered structures in water which present polar regions to solvent and bury non-polar regions. As with proteins, nucleic acids fold into structures containing helical regions. In these structures the sugar phosphate backbone of nucleic acids interacts effectively with water while the low-polarity heterocyclic bases (Fig. 9.3) are buried. The structure of several members one class of RNA, tRNA, have been determined in detail by X-ray crystallography, and one example is shown in Fig. 9.9.

The structure of a tRNA molecule (Fig. 9.9) reveals several interesting structural features. As with globular proteins, local helical regions are linked by turns to allow this molecule to fold into a compact structure. The sugar phosphate backbone runs like a ribbon over the surface of the molecule with most of the bases encapsulated within. Almost all of the bases (71 of 76) are stacked on top of one another (Fig. 9.10). This arrangement allows the non-polar faces of the bases to make intimate contact with each other and avoid contact with water.

9.7 RNA features helical structures

The tRNA structure (Fig. 9.9) resembles an 'L' shape, being made up of two short helical regions connected by a hinge. Each helical segment consists of two portions of the RNA chain running in opposite directions,

Non-polar faces in intimate contact

Fig. 9.10 An example of the stacking of bases in RNA: the 3′ end of tRNA (highlighted in Fig. 9.9).

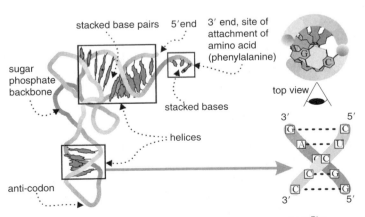

Fig. 9.9 Schematic view of the structure of yeast tRNA^{Phe}.

with each step of the helix consisting of a purine and a pyrimidine. These purine–pyrimidine pairs interact in a particular way, known as Watson–Crick base pairing (Fig. 9.11). The overall geometries of these two pairs are very similar and allow the formation of regular secondary structure. Purine–purine and pyrimidine–pyrimidine pairs cannot be accommodated within this regular structure. Hydrogen bonding takes place between the edges of the six-membered ring of each partner and allows only two discrete combinations in which the hydrogen bonding properties of the bases are complementary: adenosine pairs with uridine (A–U pair) and guanosine pairs with cytidine (G–C pair).

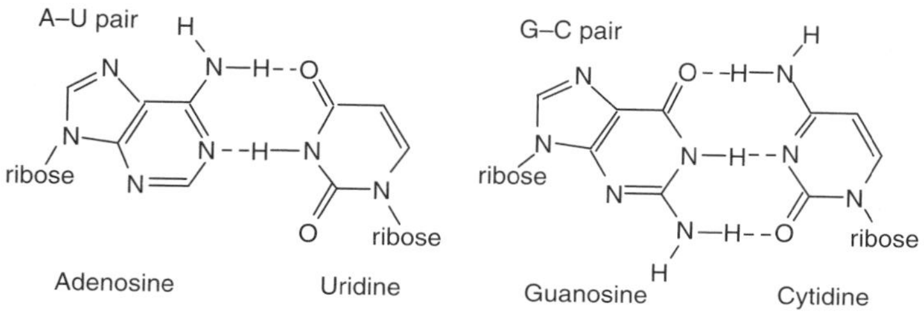

Fig. 9.11 Watson–Crick base pairing.

Steric constraints favour purine–pyrimidine pairs and specific hydrogen bonding determines the particular combinations observed. The chemical basis for these preferences is considered in detail below.

9.8 Hydrogen bonding properties of nucleotide bases

It is necessary to consider the hydrogen bonding properties of the heterocyclic bases in order to understand the specific way in which genetic information is stored and transmitted. The pairs of purines and of pyrimidines present in nucleotides differ in the pattern of nitrogen and oxygen substituents around the ring.

The information content of nucleic acids ultimately resides in a small and subtle molecular distinction: the different chemistry, including hydrogen bonding properties, of an oxygen compared with a nitrogen substituent on the heterocyclic rings. Figure 9.12 illustrates the complementary hydrogen bonding patterns of the two Watson–Crick base pairs, and relates these to the location of nitrogen and oxygen substituents.

The location of oxygen and nitrogen in the six-membered rings affects the chemical properties of the heterocyclic structure. In all cases, the oxygen or nitrogen substituent is attached to a carbon which is also attached to a nitrogen in the ring. When the substituent is oxygen, there are two possible structures (*tautomers*) depending on whether the proton associated with the chemical functional group lies on oxygen or on nitrogen (Fig. 9.13). In one tautomer, the proton is associated with the nitrogen, whereas in the other it is bonded to the exocyclic oxygen. The former structure is lower in energy under normal conditions and the 'keto' form predominates.

The adoption of helical structures by proteins was discussed in Chapter 3. The structures of nucleic acids provide another variation on this theme. The β-sheet structure of proteins (Section 3.3) is somewhat analogous to the situation observed in tRNA; the polymeric chain folds back to allow hydrogen bonding between the edges of two strands. Another analogy is with the coiled-coil structure of α-keratin (Section 3.4) where two helical strands, each with a hydrophobic face, intertwine to form a double-helical rope structure. This analogy is even more striking in the double-helical structure of DNA (Section 9.9) where two separate chains associate.

Watson–Crick base pairing was first postulated by Watson and Crick when they proposed the double-helical structure of DNA (see Section 9.9).
The bases preferentially hydrogen bond to one another, rather than to water, because it is energetically favourable for their non-polar faces to be buried away from the aqueous environment (Fig. 9.3).

A comparison has been drawn between lipids and nucleotides. One contrast between the fatty acids of lipids and the bases of nucleotides is the range of functional groups present in the latter. A key feature of RNA is that both the heterocyclic bases and the 2'-OH group are capable of hydrogen bonding. This chemical potential is exploited in cells. It is now known that some RNA molecules fold into well-defined three-dimensional shapes capable of catalysing reactions. This behaviour, analogous to that of enzymes, has led to these RNA molecules being dubbed ribozymes. Catalysis by RNA is important in a range of biochemical processes involving RNA, including the operation of the ribosome, the molecular factory which produces proteins (Sections 7.2.2 and 9.11).

An exocyclic group is one that lies outside the ring, as opposed to an endocyclic one that is incorporated as part of the covalent structure of the ring.

In guanosine there is an extra exocyclic amino group on the six-membered ring. This allows it to form an extra hydrogen bond with the pyrimidine base (see Fig. 9.11). This stabilizes the G–C base pair, but is not responsible for the selectivity of base pairing, since both pyrimidines have a keto group at the complementary position.

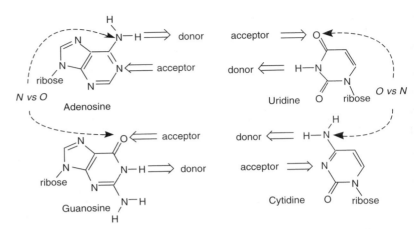

Fig. 9.12 Summary of hydrogen bonding properties related to the relative positions of oxygen and nitrogen substituents.

Structures that have the same pattern of bonds except for the location of protons which can be exchanged relatively readily, are known as 'tautomers'. Since their interconversion is easier than most other kinds of isomerism, it is given a special name: 'tautomerism'.

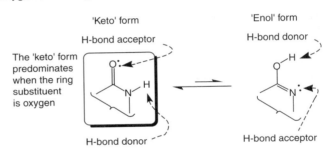

Fig. 9.13 Tautomers of oxygen-substituted heterocyclic bases.

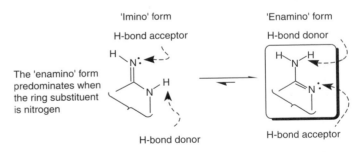

Fig. 9.14 Tautomers of nitrogen-substituted heterocyclic bases.

When the substituent is nitrogen, there is a total of two protons bonded to the relevant nitrogen atoms; again there are two possible ways of arranging these protons (Fig. 9.14). In one tautomer, the 'imino' form (analogous to the 'keto' tautomer described above), each nitrogen has a proton attached, whereas in the other, the 'enamino' form, the exocyclic nitrogen has both protons and the ring nitrogen has none. The latter structure is lower in energy under normal conditions; the 'enamino' form predominates.

Thus the change of nitrogen to oxygen (and vice versa) changes the nature of the preferred tautomer. This, in turn, changes a hydrogen bond donor–acceptor combination to an acceptor–donor arrangement. This apparently innocuous difference is a key factor underlying the mechanism by which DNA encodes genetic information.

9.9 Double-stranded nucleic acids and genetic information

The structure of tRNA demonstrates that appropriate nucleic acid sequences running in opposite directions can form stable helical structures with complementary hydrogen bonding patterns. In this case a tightly packed structure can be generated by folding the chain in a manner analogous to that of the polypeptide chains in globular proteins. The same structural preference can stabilize dimer formation between a pair of nucleic acid chains. If two discrete nucleic acid chains running in opposite directions (antiparallel) bear complementary sequences of bases, then a stable helical structure results—a double helix. In such a structure, both chains independently carry information. It is this complementarity which lies at the heart of the transfer of genetic information. These duplexes can be formed by RNA, by DNA, and by a 1 : 1 mixture of RNA and DNA. Representative duplexes are shown in Fig. 9.15.

Duplex is a term often used to describe an arrangement in which two complementary nucleic acid chains are associated.

The generation of complementary strands of nucleic acids as a means of duplicating and manipulating genetic information will be discussed further in Section 9.11.

The overall features of duplexes formed by complementary chains of RNA and DNA are the same; they both adopt antiparallel helical structures. The exact details differ because of slight differences in the conformational preferences of ribonucleotide and deoxyribonucleotide residues.

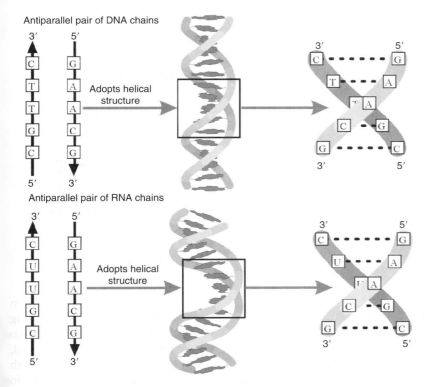

Fig. 9.15 Examples of double helices formed by RNA and DNA.

2-Deoxyribonucleotides are generated from ribonucleotides by enzyme-mediated reduction. Some people view this as evidence that 2-deoxy-ribonucleotides are specialized adaptations of ribonucleotides.

Internal attack by the hydroxyl group at C-2′ leads to the generation of a cyclic phosphate diester (Fig. 9.16). This, in turn, hydrolyses further to generate an acyclic phosphate monoester.

The other difference between DNA and RNA is in the identity of one of the pyrimidine bases: T vs U. This also reflects the need for accuracy in maintaining substantial stores of genetic information. C derivatives undergo slow hydrolysis to U analogues (see below). This process damages the reliability of genetic information: U has the opposite hydrogen bonding pattern to C, and so A rather than G would be introduced into a complementary strand. By adding a methyl group to U, this problem can be minimized: T has the same hydrogen bonding properties as U, but the two can be distinguished. 'Editing' enzymes patrol DNA and any nucleotides containing U, rather than T, can be identified as errors and then repaired.

In cells, RNA acts as a 'transient' form of information; errors are not transmitted to the next generation and, with fidelity less important, U suffices.

9.10 DNA: the genetic information of cells

As has been mentioned previously, RNA is used in a variety of roles in biology. By contrast, DNA is the polymer associated with genetic inheritance. Why is DNA so suitable for this particular role? The two polymers are distinguished by the nature of the sugar utilized in the backbone. In DNA the sugar is 2-deoxyribose rather than ribose itself. This apparently minor modification in structure has a dramatic effect on the chemical stability of the polymer and underpins the adoption of DNA as the molecule of genetic inheritance of cells.

9.10.1 The relative stability of RNA and DNA

RNA is hydrolysed rather readily in water as a result of acid or, particularly, base catalysis. Phosphate esters in general are relatively slow to hydrolyse because of electrostatic repulsion between the anionic phosphate group and the incoming nucleophile. In RNA, the hydroxyl group at C-2′ of the ring is held in close proximity to the phosphate linkage, providing a specific mechanism for nucleophilic attack (Fig. 9.16).

By removing the hydroxyl group on C-2′, this facile mechanism for cleavage of the phosphate backbone of the polymer is eliminated (Fig. 9.17). DNA is, therefore, less prone to hydrolysis in aqueous solution. This increased stability is of benefit in maintaining intact the substantial nucleic acid chains that carry genetic information for whole organisms.

Hydrolytic instability is not as great a problem for a molecule acting as a 'transient' form of information as it is for a molecule acting as the 'permanent' genetic information of a complex organism. RNA is well suited to this type of role, and it can readily be recycled once its job is done.

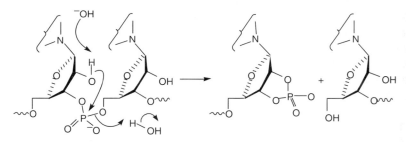

Fig. 9.16 Base-catalysed hydrolysis of the backbone of RNA.

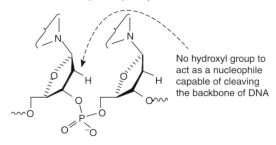

No hydroxyl group to act as a nucleophile capable of cleaving the backbone of DNA

Fig. 9.17 The absence of a 2′-hydroxyl group in DNA increases hydrolytic stability.

9.11 The flow of information in cells

This section outlines the three main processes of information transfer in cells. DNA *replication*, the production of an identical copy of DNA during cell division, occurs in order to pass genetic information to future generations of cells. *Transcription* is the process whereby the specific information contained within DNA is read by making an RNA complement of a selected segment of DNA. Some of this RNA, mRNA, carries the information required for protein synthesis. This information is decoded to allow the production of proteins of defined sequence, a process known as *translation*.

9.11.1 Nucleic acid biosynthesis

In the double-helical structures of nucleic acids (Fig. 9.15), each individual strand carries all the information required for the generation of another copy of the other strand. Given one strand, a complementary strand can be produced by placing G opposite C and A opposite U or T. Likewise, a strand of one nucleic acid can be used as a template to generate a complementary strand of a different type.

The chemistries of DNA and RNA polymerization are closely related (Fig. 9.18). In each case, a growing chain is hydrogen bonded to a template strand. A nucleoside triphosphate, complementary to that on the template chain, binds at the next vacant site. An enzyme catalyses the nucleophilic attack by the 3′hydroxyl group of the growing chain at the α-phosphate of the monomer to generate the phosphate ester linkage.

9.11.2 DNA replication: the basis of genetic inheritance

The permanent store of genetic information in cells is double-stranded DNA. When a cell divides, both daughter cells require a copy of the double-stranded DNA. Each daughter cell receives one of the original parental strands, whilst DNA synthesis provides a new copy of the complementary strand (Fig. 9.19). The overall effect is that two copies of the original double-stranded DNA are produced; the DNA is replicated.

Generating an RNA sequence from a selected portion of DNA can be likened to transcribing a message. Hence the description of this process as 'transcription', and the name for the resulting RNA, which carries the message for the protein, as messenger RNA (or mRNA). The information carried as the nucleic acids RNA and DNA is in a common 'language', but requires 'translation' into the language of proteins. The translating molecule which carries information in both languages, and which transfers amino acids to the growing protein chain, is known as transfer RNA, or tRNA (Fig. 9.21).

Synthesis of DNA molecules is catalysed by enzymes known as DNA polymerases. Likewise RNA polymerases mediate the synthesis of new strands of RNA.

In normal cellular activity, DNA information is converted into RNA and thence into proteins. The 'languages' of RNA and DNA are equivalent and it is possible to use RNA as a template to make DNA. This process, known as 'reverse transcription', is a feature of some viruses which use RNA as their principal information store. This type of virus includes HIV, responsible for AIDS (viruses are discussed further in an aside on the next page).

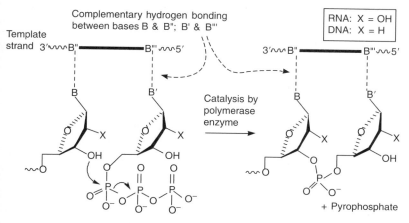

Fig. 9.18 The biosynthesis of nucleic acids.

Viruses cannot reproduce by themselves; instead they invade cells and hijack cellular enzymes and metabolites to produce more viral material. A virus is comprised of a segment of RNA or DNA encapsulated in a 'coat' made from protein molecules, the sequences of which are encoded by the viral nucleic acid. Viruses have much less genetic information than other organisms and so they do not require this information to be as stable—in fact, they benefit from errors to the extent that these increase the rate of evolution.

Although information can be interconverted between RNA and DNA, the flow of information into proteins is one way—proteins cannot be used as templates for nucleic acid generation. This asymmetry between the information content of nucleic acids and proteins is often termed the 'central dogma' of molecular biology.

In a region of double-stranded DNA associated with a gene (a functional unit of genetic information that typically codes for a specific protein), one strand corresponds to the gene and the other, a 'template' strand, is its complement. The mRNA is constructed on the template strand and is, therefore, effectively a copy of the 'informational' strand (noting, of course, the change in sugar and the presence of U, rather than T, residues). This is shown in Fig. 9.22.

Because there are 64 codons and 21 items to encode (20 amino acids and a stop signal), there are redundancies in the code. It can be seen from Fig. 9.20 that most of these redundancies are associated with the third position of the triplet. For many amino acids, the first two positions of the codon are specific, but some changes can be tolerated at the third site.

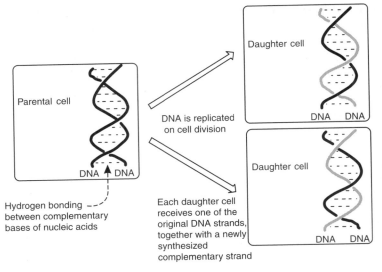

Fig. 9.19 Schematic view of DNA replication occurring on cell division.

9.11.3 Transcription and translation: proteins on demand

Individual portions of the DNA molecule provide the information for the construction of various RNA molecules and proteins. RNA corresponding to a region of interest is produced by transcription—the process whereby an RNA strand is synthesized from a DNA template (see Fig. 9.18). The information for protein synthesis is not accessed directly. Instead it resides in one of the classes of RNA, 'messenger' RNA ('m-RNA'), produced by transcription.

Messenger RNA is used to direct the production of protein. It carries the information required for the production of a particular protein to the ribosomes where protein biosynthesis occurs. The relationship between the linear sequence of an mRNA molecule and the linear sequence of the corresponding protein is known as the 'genetic code'. There are four types of nucleotide, but 20 amino acid building blocks in proteins. There cannot, therefore, be a 1:1 correspondence between the nucleotide order and the amino acid order. Instead, the RNA is read in triplets: three contiguous nucleotides specify a single amino acid. There are 64 (4^3) such 'codons' (Fig. 9.20).

Amino acid/codons		Amino acid/codons	Amino acid/codons
Arg	CGU, CGC, CGA, CGG, AGA, AGG	Ile AUU, AUC, AUA	Gln CAA, CAC
Leu	UUA, UUG, CUU, CUC, CUA, CUG	Asn AAU, AAC	Glu GAA, GAG
Ser	UCU, UCC, UCA, UCG, AGU, AGC	Asp GAU, GAC	Lys AAA, AAG
Ala	GCU, GCC, GCA, GCG	Cys UGU, UGC	Phe UUU, UUC
Gly	GGU, GGC, GGA, GGG	His CAU, CAC	Tyr UAU, UAC
Pro	CCU, CCC, CCA, CCG		Trp UGG
Thr	ACU, ACC, ACA, ACG		
Val	GUU, GUC, GUA, GUG	Stop UAA, UGA, UAG	Met (*Start*) AUG

Fig. 9.20 The genetic code.

Transfer RNA acts as an intermediary in the translation of mRNA into protein, recognizing triplets ('reading the code') and specifying the amino acid to be incorporated. There is a family of tRNA molecules, each member specifying a particular amino acid. As part of its structure, each tRNA contains a segment called the anti-codon (Fig. 9.9), a sequence of three ribonucleotides which form complementary hydrogen bonds to the triplet codon sequence. The tRNA bears the amino acid specified by this codon sequence, bound by an ester link to the 3′-hydroxyl of the 3′-terminal ribose.

The codons on the mRNA are read in sequence. Formation of complementary hydrogen bonds between the set of three ribonucleotides of a codon on the mRNA, and those comprising the anti-codon of the tRNA, bring the appropriate amino acid into position for enzyme-mediated transfer to the growing protein chain (Fig. 9.21).

An overview of transcription and translation is given in Fig. 9.22. RNA polymerase, the enzyme which catalyses transcription of DNA into RNA, binds at a region of DNA called the promoter. Then, the RNA polymerase catalyses the generation of an RNA copy of the succeeding sequence of DNA. The binding of RNA polymerase can be controlled to regulate the amount of mRNA, and ultimately protein, produced.

Each amino acid is loaded onto the correct tRNA by a specific enzyme, an amino acyl tRNA synthetase. Each of these enzymes recognizes both a specific amino acid and the correct tRNA. These enzymes, therefore, act as the discriminators that recognize the languages of both nucleic acids and proteins by identifying characteristic features of the tRNA and the corresponding amino acid.

Section 7.2 discussed the chemistry of protein biosynthesis.

A gene typically has three regions: a promoter where RNA polymerase binds before transcribing the subsequent nucleotide sequence; the coding region, which encodes a particular protein; and finally a further specific base sequence which signals the end of a gene.

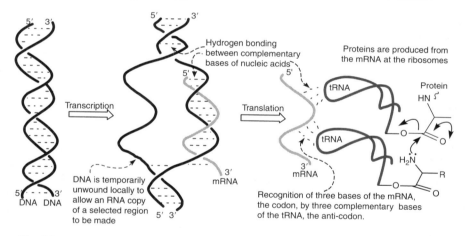

Fig. 9.21 Schematic view of the flow of information between DNA, RNA and protein.

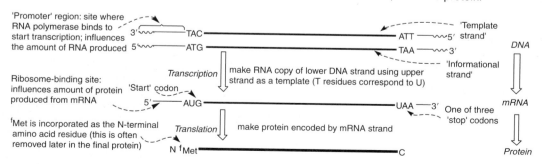

Fig. 9.22 Schematic overview of transcription and translation.

The amount of a protein produced (the level of 'gene expression') is regulated by controlling the activity of the relevant polymerase enzymes. Some proteins, e.g., bind at the promoter region and prevent RNA polymerase binding, there by decreasing the amount of mRNA and, consequently, protein generated. Conversely, proteins that bind to DNA and stimulate RNA polymerase binding can increase the amount of mRNA, and hence protein, synthesized.

The first residue of a growing protein chain is *N*-formyl methionine (f) Met in which the amino group is masked:

AUG codons have a dual role: specifying incorporation of *N*-formyl methionine at the start point and 'normal' methionine within the coding region itself.

A few important genetic engineering techniques are listed below, to provide a glimpse of the methodologies that have been developed. DNA can be cut at specific sites with enzymes called restriction enzymes. DNA fragments from different sources, cut by a particular enzyme, can be stuck together ('ligated') by the use of enzymes called ligases. DNA is a charged molecule and will not enter cells without assistance; methods have been developed which allow DNA to be introduced into cells, e.g. 'transformation' of *E. coli* is usually facilitated by bathing the cells in aqueous $CaCl_2$. DNA sequencing has now become so efficient that the ordering of nucleotides in complete genomes (billions of bases in a unique sequence) can be determined. Finally, preparation of synthetic DNA is now so sophisticated that it can be accomplished efficiently by machine.

The generation of protein from RNA takes place at ribosomes, a complex assembly of proteins and RNA. Ribosomes bind at the 5′ end of mRNA (at the 'ribosome-binding site'). The presence of a nearby AUG triplet indicates a start site of the 'coding region'. A new protein chain is started at this point using *N*-formyl methionine as the first amino acid. This identifies the 'reading frame' since succeeding triplets are read in register. The subsequent presence of one of three 'stop codons' specifies the point at which construction of the protein chain ends. The completed protein is then released from the ribosome.

9.12 Genetic engineering

Cells have the ability to replicate their DNA and produce proteins via the processes of transcription and translation. Foreign DNA entering a cell can hijack this replicative ability, provided it contains the information required for the appropriate cellular enzymes to act. For example, viruses infect cells and use cellular nucleic acid polymerases and ribosomes to generate more viral nucleic acid and proteins which then self-assemble to produce new viruses.

This replicative ability is also exploited in the field of 'genetic engineering'. Genetic engineering involves the artificial generation of modified organisms by changing the DNA content of cells (Fig. 9.23).

Introduction of DNA that encodes a protein can lead to a new organism that produces this extra protein. The new organism may be of interest in its own right (e.g. the protein might correct a genetic defect in

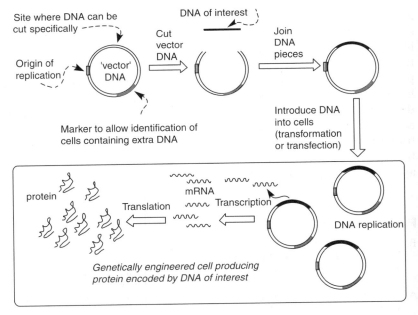

Fig. 9.23 Stylized overview of genetic engineering.

the original host, the basis of 'gene therapy') or may simply be a useful source of the new protein. Genetic engineering experiments depend on the ability to isolate and characterize DNA of interest and combine it with DNA that allows the new sequence to be replicated in the chosen host organism. It is necessary to be able to hydrolyse DNA at specific sites, join pieces of DNA from different sources, introduce DNA into cells, and identify and grow cells that contain the extra DNA. The ability to sequence and synthesize DNA chemically is also a very valuable tool. Extensive research during the latter stages of the twentieth century established this methodology, and genetic engineering has become a routine scientific technique.

Genetically engineered microorganisms can be used to produce large amounts of proteins for many purposes, e.g. human insulin is now produced for the treatment of diabetes using this methodology. Proteins can also be manipulated in a systematic way. By changing the DNA that encodes a given protein, the protein sequence can be altered. As an example, Section 5.7 described how changing specific amino acids in an enzyme (TIM) has been used to provide detailed information about catalysis.

A DNA molecule that can be used to replicate an attached piece of DNA is called a vector. The vector DNA shown in Fig. 9.23 is a 'plasmid': circular DNA, which can replicate in a host organism by virtue of having an 'origin of replication' (a site recognized by host DNA replication enzymes). Vector DNA also includes one or more 'markers' which allow identification of cells that have taken up plasmid DNA. Bacteria, such as *E. coli*, are common host organisms and resistance to antibiotics is an important type of marker. For example, *E. coli* cannot grow in the presence of significant amounts of penicillin (see Section 5.10). Enzymes, known as β-lactamases, which catalyse the destruction of penicillins, allow *E. coli* to grow in penicillin-containing media. DNA which encodes β-lactamase can be used as a marker: if cells produced by a genetic engineering experiment are grown in penicillin-containing media, only cells which have taken up plasmid DNA (containing the β-lactamase gene and the gene of interest) will grow.

In an analogous fashion, genes encoding resistance to herbicides can be introduced into plants. This technology has been used to produce genetically modified food-producing plants. When harvested these plants give rise to 'GM' foods whose production and availability has caused considerable debate.

9.13 Summary

Nucleotides are heterocyclic bases linked to a sugar phosphate residue, derived from either ribose or 2-deoxyribose. These units can polymerize to give nucleic acids by ester formation between an alcohol of one nucleotide and a phosphate of another. The sugar phosphate backbone of a nucleic acid is polar, whilst the bases are less polar. In aqueous solution, nucleic acids adopt ordered structures in which the sugar phosphate groups are in contact with water and the bases are buried. An important feature is that helices can form between nucleic acid strands running in opposite directions. These strands interact through complementary hydrogen bonding patterns between bases, determined by the substitution patterns on the heterocyclic bases. This complementary hydrogen bonding is the basis of the storage and transmission of genetic information. It is used in duplicating DNA, the permanent information store of the organism, and in transcription to form RNA, which is used to generate proteins via translation. Genetic engineering exploits biological replication to produce and manipulate proteins of interest.

Further reading

Biochemistry textbooks, such as those listed in Chapter 1, discuss nucleic acid biochemistry, e.g. C. K. Mathews, K. E. van Holde and K. G. Ahern (2000) *Biochemistry*, 3rd edn, Benjamin/Cummings, San Francisco, Part V; an overview of genetic engineering techniques is given in pp. 969–79.

An overview of genetic engineering is given in J. D. Watson, M. Gilman, J. Witkowski and M. Zoller (1992) *Recombinant DNA*, 2nd edn, W. H. Freeman, Basingstoke.

Scientific American is a useful starting point for finding out about the science behind some contemporary issues in genetic engineering. Some recent topics include: research on transgenic animals, *Sci. Am.*, Dec. 1998, pp 30–35; and human genome research, *Sci. Am.*, Jul. 2000, pp. 38–57.

Wider human genetic engineering issues are addressed in K. A. Drlica (1994) *Double-Edged Sword*, Helix Books, Reading, Massachusetts.

As an example of a contemporary genetic engineering issue, the pros and cons of genetically engineered food are discussed in A. McHughen (2000) *A Consumers Guide to GM Food*, Oxford University Press, Oxford.

10 Epilogue: where to from here?

This book has introduced some of the intrinsic chemical properties of important types of biological molecules, especially amino acids, sugar phosphate derivatives and related macromolecules (proteins and nucleic acids). It has shown how a small number of fundamental chemical properties (particularly acid–base chemistry, electrostatic charge, hydrogen bonding and stereochemistry) can be used to understand many of the biological functions of these molecules. As well as discussing the chemical potential of the functional groups found in biological molecules, this book has highlighted the way in which amphiphilic molecules, e.g. macromolecules and lipids, can adopt ordered, functionalised structures in an aqueous environment.

The availability of such ordered structures with diverse chemical functional groups provides the foundations for the molecular basis of life. In particular, macromolecules that adopt extended structures are used by biology as fibres (proteins and polysaccharides) and for information storage (DNA); proteins that fold into compact structures are used to bind other molecules for the purposes of transport (proteins) and catalysis (proteins and RNA). In addition, structures built from macromolecules and from assemblies of phospholipids provide crucial interfaces between different biological compartments, and between their compartments and the outside world.

We hope that this book provides a good starting point for understanding chemical biology. The best way to further your understanding is by solving problems, e.g. answering questions found in textbooks. But science really is about developing a questioning attitude to the world and you can generate your own problems by questioning what you read and see about you. For example, when you encounter a metabolic reaction for the first time, you might ask: is this like any metabolic reaction I have seen before? What are the possible organic chemical mechanisms for the process? How might an enzyme catalyse the reaction? If a coenzyme is needed, what is its chemical role? What is the role of the process for the organism that undertakes it?

As you study further, you will see the basic concepts of chemical biology applied to ever more sophisticated systems. You will encounter more types of naturally occurring molecule, more metabolic reactions, more complex macromolecular structures and the use of more chemical elements. All these chemical processes are organized and regulated in a sophisticated manner. The basic chemical principles, however, remain the same as those that you will have read about in this book.

The ability to self-assemble into complex structures is one of the most remarkable and important features of the molecules within living systems.

The same factors that are exploited by nature to produce molecules with interesting properties are increasingly used in other areas of chemistry. As an example, synthetic organic materials, such as polyesters, are now produced with sophisticated properties tailored to particular needs. It is informative to apply and extend the knowledge gained from this book to such related areas of science. As well as understanding more about the world we live in, this will also consolidate your understanding of chemical biology.

Chemical biology has many exciting applications which are also the focus of a great deal of research. Important areas include medical diagnosis and treatment (new drugs and therapies), environmental science, materials science and forensic science (e.g. DNA fingerprinting).

Furthermore, you will find that at the frontiers of chemical biology there is a vast range of exciting problems that challenge the ingenuity of research scientists. How can we make and modify complex molecules found in nature? How do proteins fold to form complex structures? Can we predict such structures from amino acid sequences? How can we understand the biological functions of proteins identified by the sequencing of the human genome? How can we study individual biological molecules in action? What chemistry can we study as it takes place in living cells? And finally, can chemical, a subject that promises so much for our future health and prosperity, even provide a molecular glimpse into the past to understand how life began on primitive earth?

The answers to these and a myriad of other equally fascinating and important questions will undoubtedly emerge in the future, as research in this interdisciplinary area continues to attract scientists from a wide variety of traditional disciplines. We hope that this short book has excited your interest to the extent that you will follow the future developments in chemical biology, and perhaps might even have stimulated you to take part in some of the many research opportunities and challenges that lie ahead.

Index